NANOTECHNOLOGY SCIENCE AND TECHNOLOGY

NANOTECHNOLOGY CONSIDERATIONS FOR THE EPA AND FDA

NANOTECHNOLOGY SCIENCE AND TECHNOLOGY

Additional books in this series can be found on Nova's website under the Series tab.

Additional E-books in this series can be found on Nova's website under the E-book tab.

NANOTECHNOLOGY SCIENCE AND TECHNOLOGY

NANOTECHNOLOGY CONSIDERATIONS FOR THE EPA AND FDA

SCOTT J. CASTIGLIONE
EDITOR

Nova Science Publishers, Inc.
New York

For permission to use material from this book please contact us:
Telephone 631-231-7269; Fax 631-231-8175
Web Site: http://www.novapublishers.com

LIBRARY OF CONGRESS CATALOGING-IN-PUBLICATION DATA

Nanotechnology considerations for the EPA and FDA / editor, Scott J. Castiglione.
p. cm.
Includes index.
ISBN 978-1-61761-817-8 (hardcover)
1. Nanoparticles--Environonmental aspects. 2. Nanoparticles--Toxicology. 3. Nanostructured materials--Environmental aspects. 4. Nanostructured materials--Health aspects. 5. Nanotechnology--Environmental aspects. 6. Nanotechnology--Health aspects. 7. Nanotechnology--Government policy--United States. I. Castiglione, Scott J.
TD196.N36N37 2010
363.17--dc22

2010033082

Published by Nova Science Publishers, Inc. † New York

CONTENTS

PREFACE

Nanotechnology, a term encompassing nanoscale science, engineering and technology, is focused on understanding, controlling, and exploiting the unique properties of matter than can emerge at scales of 1 to 100 nanometers. A key issue before Congress regarding nanotechnology is how best to protect human health, safety and the environment as nanoscale materials and products are researched, developed, manufactured, used, and discarded. While the rapidly emerging field of nanotechnology is believed by many to offer significant economic and societal benefits, some research results have raised concerns about the potential adverse environmental, health, and safety implications of nanoscale materials. This book explores the nanotechnology considerations facing the EPA and FDA today

Chapter 1- Nanotechnology involves the ability to control matter at the scale of a nanometer—one billionth of a meter. The world market for products that contain nanomaterials is expected to reach $2.6 trillion by 2015. In this context, GAO (1) identified examples of current and potential uses of nanomaterials, (2) determined what is known about the potential human health and environmental risks from nanomaterials, (3) assessed actions EPA has taken to better understand and regulate the risks posed by nanomaterials as well as its authorities to do so, and (4) identified approaches that other selected national authorities and actions U.S. states have taken to address the potential risks associated with nanomaterials. GAO analyzed selected laws and regulations, reviewed information on EPA's Nanoscale Materials Stewardship Program, and consulted with EPA officials and legal experts to obtain their perspectives on EPA's authorities to regulate nanomaterials.

Chapter 2- Nanotechnology—a term encompassing nanoscale science, engineering, and technology—is focused on understanding, controlling, and exploiting the unique properties of matter that can emerge at scales of one to 100 nanometers. A key issue before Congress regarding nanotechnology is how best to protect human health, safety, and the environment as nanoscale materials and products are researched, developed, manufactured, used, and discarded. While the rapidly emerging field of nanotechnology is believed by many to offer significant economic and societal benefits, some research results have raised concerns about the potential adverse environmental, health, and safety (EHS) implications of nanoscale materials.

Some have described nanotechnology as a two-edged sword. On the one hand, some are concerned that nanoscale particles may enter and accumulate in vital organs, such as the lungs and brains, potentially causing harm or death to humans and animals, and that the diffusion of nanoscale particles in the environment might harm ecosystems. On the other hand, some believe that nanotechnology has the potential to deliver important EHS benefits such as reducing energy consumption, pollution, and greenhouse gas emissions; remediating environmental damage; curing, managing, or preventing diseases; and offering new safety-enhancing materials that are stronger, self-repairing, and able to adapt to provide protection.

Chapter 3- As other emerging technologies have in the past, nanotechnology poses questions regarding the adequacy and application of regulatory authorities. The then Acting Commissioner of the Food and Drug Administration (FDA) initiated the Nanotechnology Task Force (Task Force) in 2006 to help assess these questions with respect to FDA's regulatory authorities, in light of the current state of the science for nanotechnology. This report offers the Task Force's initial findings and recommendations to the Commissioner.

The report includes:

- A synopsis of the state of the science for biological interactions of nanoscale materials;
- Analysis and recommendations for science issues; and
- Analysis and recommendations for regulatory policy issues.

The report addresses scientific issues as distinct from regulatory policy issues in recognition of the important role of the science in developing regulatory policies in this area, rapid growth of the field of nanotechnology, and the evolving state of scientific knowledge relating to this field. Rapid developments in the field mean that attention to the emerging science is needed to enable the agency to predict and prepare for the types of products FDA may see in the near future.

In: Nanotechnology Considerations for the EPA... ISBN: 978-1-61761-817-8
Editors: Scott J. Castiglione

Chapter 1

NANOTECHNOLOGY: NANOMATERIALS ARE WIDELY USED IN COMMERCE, BUT EPA FACES CHALLENGES IN REGULATING RISK

United States Government Accountability Office

WHY GAO DID THIS STUDY

Nanotechnology involves the ability to control matter at the scale of a nanometer—one billionth of a meter. The world market for products that contain nanomaterials is expected to reach $2.6 trillion by 2015. In this context, GAO (1) identified examples of current and potential uses of nanomaterials, (2) determined what is known about the potential human health and environmental risks from nanomaterials, (3) assessed actions EPA has taken to better understand and regulate the risks posed by nanomaterials as well as its authorities to do so, and (4) identified approaches that other selected national authorities and actions U.S. states have taken to address the potential risks associated with nanomaterials. GAO analyzed selected laws and regulations, reviewed information on EPA's Nanoscale Materials Stewardship Program, and consulted with EPA officials and legal experts to obtain their perspectives on EPA's authorities to regulate nanomaterials.

What GAO Recommends

GAO recommends that EPA complete its plans to modify its regulatory framework for nanomaterials as needed. EPA concurred with our recommendations and provided technical comments, which we incorporated as appropriate.

What GAO Found

Companies around the world are currently harnessing the properties of nanomaterials for use in products across a number of sectors and are expected to continue to find new uses for these materials. GAO identified a variety of products that currently incorporate nanomaterials already available in commerce across the following eight sectors: automotive; defense and aerospace; electronics and computers; energy and environment; food and agriculture; housing and construction; medical and pharmaceutical; and personal care, cosmetics and other consumer products. Within each of these sectors, GAO also identified a wide variety of other uses that are currently under development and are expected to be available in the future.

The extent to which nanomaterials present a risk to human health and the environment depends on a combination of the toxicity of specific nanomaterials and the route and level of exposure to these materials. Although the body of research related to nanomaterials is growing, the current understanding of the risks posed by these materials is limited. This is because the manner in which some studies have been conducted does not allow for valid comparisons with newer studies or because there has been a greater focus on certain nanomaterials and not others. Moreover, the ability to conduct necessary research on the toxicity and risks of nanomaterials may be further hampered by the lack of tools to conduct such studies and the lack of models to predict the characteristics of nanomaterials.

EPA has undertaken a multipronged approach to understanding and regulating the risks of nanomaterials, including conducting research and implementing a voluntary data collection program. Furthermore, under its existing statutory framework, EPA has regulated some nanomaterials but not others. Although EPA is planning to issue additional regulations later this year, these changes have not yet gone into effect and products may be entering the market without EPA review of all available information on their potential risk. Moreover, EPA faces challenges in effectively regulating nanomaterials that may be released in air, water, and waste because it lacks the technology to monitor and characterize these materials or the statutes include volume based regulatory thresholds that may be too high for effectively regulating the production and disposal of nanomaterials.

Like the United States, Australia, Canada, the United Kingdom, and the European Union have begun collecting data to understand the potential risks associated with nanomaterials and are reviewing their legislative authorities to determine the need for modifications. Australia and the United Kingdom have undertaken a voluntary data collection approach whereas Canada plans to require companies to submit certain types of information. Some U.S. states, like California, have also begun to address the potential risks from nanomaterials by, for example, collecting information from manufacturers on a limited number of nanomaterials in use in those states and making some of this information publicly available.

ABBREVIATIONS

CERCLA	Comprehensive Environmental Response, Compensation, and Liability Act
EPA	Environmental Protection Agency
FIFRA	Federal Insecticide, Fungicide, and Rodenticide Act
ISO	International Organization for Standardization
NICNAS	National Industrial Chemicals Notification and Assessment Scheme
NNI	National Nanotechnology Initiative
OECD	Organisation for Economic Co-operation and Development
RCRA	Resource Conservation and Recovery Act
REACH	Regulation, Evaluation and Authorization of Chemicals
SNUR	Significant New Use Rule
TSCA	Toxic Substances Control Act of 1976
UV	ultraviolet
Wilson Center	Woodrow Wilson International Center for Scholars' Project on Emerging Nanotechnologies

May 25, 2010
The Honorable Barbara Boxer
Chairman
Committee on Environment and Public Works
United States Senate

Dear Madam Chairman:

The term "nanotechnology" encompasses a wide range of innovations based on the understanding and control of matter at the scale of nanometers—the equivalent of one-billionth of a meter. For illustration, a sheet of paper is about 100,000 nanometers thick, a human hair is about 80,000 nanometers wide, and three gold atoms lying side by side are about 1 nanometer long. Unusual properties can emerge in materials manufactured at the nanoscale—including catalytic, electrical, magnetic, mechanical, optical, and thermal properties—that differ in important ways from the properties of conventionally scaled materials. Some of these new properties can enhance products and their applications across a number of sectors, including electronics, medicine, and defense. The world market for nanotechnology-related products is growing and is expected to total between $1 trillion and $2.6 trillion by 2015.

Nanomaterials can occur naturally, be created incidentally, or be manufactured intentionally. For example, naturally occurring nanomaterials can be found in volcanic ash, forest fire smoke, and ocean spray. Incidental nanomaterials are by-products of industrial processes, such as mining and metal working, and combustion engines, such as those used in cars, trucks, and some trains. In contrast, manufactured nanomaterials (sometimes called engineered nanomaterials) have been specifically designed for a particular function or property, such as improved strength, decreased weight, or increased electrical conductivity. Our review will focus on manufactured nanomaterials, rather than nano-sized materials that occur naturally in the environment or are incidentally produced, and for the remainder of this report, we will call such materials "manufactured nanomaterials," or simply "nanomaterials." While the use of nanomaterials holds promise for the future, their small size and unique properties raise questions about potential risks to people or the environment that might result from exposure to them during their manufacture, use, and disposal. Risk is usually defined as the potential for harmful effects to human health or the environment resulting from exposure to a substance—in this case, nanomaterials. In general terms, risk depends on a combination of the exposure a person or the environment has to the substance as well as the inherent toxicity of the chemical. In other words, the same exposure to two different substances each with their own toxicity would result in different levels of potential risk.

The Environmental Protection Agency (EPA) administers several laws that regulate chemicals, pesticides, pollutants in air or water, and wastes that may be composed of or contain nanomaterials.[1] These laws include the following:

- the Toxic Substances Control Act of 1976 (TSCA), which authorizes EPA to require chemical companies to report certain information about chemicals used in commerce and authorizes EPA to require testing of and control chemicals that pose an unreasonable risk to human health or the environment, among other things;
- the Federal Insecticide, Fungicide, and Rodenticide Act (FIFRA), which authorizes EPA to regulate the sale and use of pesticides and prohibits marketing of pesticides that have not been registered with EPA;[2]
- the Clean Air Act, which requires EPA to set standards for common air pollutants and to regulate industrial sources of hazardous air pollutants;
- the Clean Water Act, which authorizes EPA to regulate discharges of pollutants into federally regulated waters;
- the Resource Conservation and Recovery Act (RCRA), which establishes a framework for regulation of hazardous and solid wastes and authorizes EPA to issue administrative orders to address imminent hazards; and
- the Comprehensive Environmental Response, Compensation, and Liability Act (CERCLA), commonly known as Superfund, which authorizes EPA to compel parties responsible for contaminating sites to clean them up or to conduct cleanups itself and then seek reimbursement from responsible parties.

On the international level, other national authorities are also concerned about the potential risks of nanomaterials and whether their current regulatory framework authorities are sufficient to address these risks. For example, Australia, Canada, the United Kingdom, and the European Union have begun to review their regulatory approaches for nanomaterials. Furthermore, the Organisation for Economic Co-operation and Development—a forum in which the governments of 30 developed countries, including the United States, work together to address economic, social, and environmental issues—has established a "working party" on nanomaterials. In addition to the international focus on this topic, some U.S. states have begun to explore ways to address the potential risks of nanomaterials.

In this context, you asked us to (1) identify examples of current and potential uses of nanomaterials, (2) determine what is known about the potential human health and environmental risks from nanomaterials, (3) specifically assess actions EPA has taken to better understand and regulate the risks posed by nanomaterials as well as its authorities to do so, and (4) identify approaches that selected other national authorities have taken to address the risks associated with nanomaterials. In addition, you asked us to identify any U.S. states and localities that have begun to address the risks from nanomaterials.

To identify examples of current and potential uses of manufactured nanomaterials, we analyzed documents and reports that discuss the current and future uses of manufactured nanomaterials, such as market research reports produced by Lux Research, an independent research firm that conducts market analysis of nanotechnology, among other things. In addition, we interviewed cognizant agency officials from the six U.S. agencies that conduct the majority of nanotechnology-related research.[3]

We also interviewed knowledgeable stakeholders, including officials from the National Nanotechnology Initiative, the Wilson Center, the National Academy of Sciences, Lux Research, and the NanoBusiness Alliance—a nanotechnology-related business association. We used an iterative process, often referred to as "snowball sampling," to identify knowledgeable stakeholders, and we selected for interviews those who would provide us with a broad range of perspectives on the current and potential uses of nanomaterials.

To determine what is known about the potential human health and environmental risks of manufactured nanomaterials, we reviewed documents that had been published by peer-reviewed journals, government agencies, and international nonprofit organizations. In conducting this review, we searched databases, asked knowledgeable stakeholders to identify relevant studies, and reviewed studies from article bibliographies to identify additional sources of information on the potential risks. Our review focused on 20 such studies, selected in part because they provided a synthesis of available research related to nanomaterials' risks and covered a variety of nanomaterials. For the purposes of this report, all the documents, studies, and synthesis studies we reviewed will be referred to as "studies." We also spoke with a variety of knowledgeable stakeholders representing industry, academia, non governmental organizations, and the regulatory community. These knowledgeable stakeholders were also selected using a snowball sampling method.

To assess actions EPA has taken to better understand and regulate manufactured nanomaterials and its authorities to do so, we analyzed selected laws and regulations, including TSCA, FIFRA, the Clean Air Act, the Clean Water Act, RCRA, and CERCLA. We also reviewed data and reports on EPA's Nanoscale Materials Stewardship Program, which EPA developed to encourage companies to voluntarily develop and submit information to the agency on the characteristics of nanomaterials. Furthermore, we consulted with EPA officials and legal experts to obtain their perspectives on EPA's available authorities to regulate manufactured nanomaterials.

To identify the approaches that other selected national authorities– Australia, Canada, the United Kingdom, and the European Union—have used to address the potential risks associated with manufactured nanomaterials, we analyzed these authorities' laws and regulations that would be applicable to regulating manufactured nanomaterials. We selected these authorities based on interviews with knowledgeable stakeholders who identified them as having taken actions related to better understanding, assessing, or regulating the potential risks of nanomaterials. To identify any states that may be taking action with regard to nanomaterials, we spoke with federal regulators; industry and environmental groups; and other knowledgeable stakeholders, including the Environmental Council of States.

A more detailed description of our scope and methodology is presented in appendix I. We performed our work between May 2009 and May 2010, in accordance with generally accepted government auditing standards. Those standards require that we plan and perform the audit to obtain sufficient, appropriate evidence to provide a reasonable basis for our findings and conclusions based on our audit objectives. We believe that the evidence obtained provides a reasonable basis for our findings and conclusions based on our audit objectives.

BACKGROUND

In fiscal year 2009, federal support for nanotechnology research totaled about $1.7 billion. Cumulatively from fiscal year 2001 through fiscal year 2009, federal agencies have devoted over $10.5 billion to nanotechnology research. To guide federal development of nanotechnology, the National Nanotechnology Initiative (NNI) was established in 2001 to support long-term research and development aimed at accelerating the discovery, development, and deployment of nanoscale science, engineering, and technology. The NNI is a mechanism to coordinate the nanotechnology-related activities of the 25 currently participating federal agencies that fund nanoscale research or have a stake in the outcome of this research, such as those agencies that may regulate products containing nanomaterials. While the NNI is designed to facilitate intergovernmental cooperation and identify overarching goals and priorities for nanotechnology research, it is not a research program and has no funding or authority to dictate the nanotechnology research agenda for participating agencies or to ensure that adequate resources are available to achieve specific goals. Instead, participating agencies develop and fund their own nanotechnology research agendas. In fiscal year 2009, six NNI agencies accounted for over 95 percent of federal nanotechnology research reported. These are the Department of Defense, the Department of Energy, EPA, the Department of Health and Human Services' National Institutes of Health, the Department of Commerce's National Institute of Standards and Technology, and the National Science Foundation.

Nanomaterials can take a variety of forms and can generally be organized into four types:

- *Carbon-based materials.* These nanomaterials are composed mostly of carbon, and are most commonly spherical, elliptical, or tubular in shape. Spherical and elliptical carbon shapes are referred to as fullerenes, while tubular ones are called nanotubes.
- *Metal-based materials*. These nanomaterials include nanoscale gold, nanoscale silver, and metal oxides, such as titanium dioxide. They also include quantum dots, which are closely packed semiconductor crystals comprised of hundreds or thousands of atoms, on the scale of a few nanometers to a few hundred nanometers.
- *Dendrimers.* These nanomaterials are nanoscale polymers built from branched units. The surface of a dendrimer has numerous branch ends, which can be tailored to perform specific chemical functions. Also, some dendrimers contain interior cavities into which other molecules can be placed, such as for drug delivery.
- *Composites.* These materials combine nanoparticles with other nanoparticles or with larger, conventional-scale materials. For

example, nanoparticles, such as nanoscale clay can be combined with other materials to form a composite material.

EPA uses a risk assessment process to estimate the extent of harm, if any, that can be expected from exposure to a given substance throughout its life cycle and to help regulators determine whether the risk meets the requirements for taking action under its statutory authorities, such as banning the substance's production or limiting its use. The basic risk assessment paradigm includes the following:

- an evaluation of scientific information on a substance's hazardous properties—or toxicity—which may potentially affect human health or the environment;
- the dose-response relationship—the relationship between the extent of exposure (dose) and the resulting changes in health or body function (response)—describes the toxic effect of a substance; and
- exposure—the extent to which humans or the environment are expected to be exposed to the chemical.

EPA is applying this risk assessment paradigm to assess the potential risks from nanomaterials. EPA officials also told us that risk assessment is not the only means of using scientific information to inform decision making. For example, they said that by using green chemistry and life cycle assessment approaches,[4] a material's properties may be modified or exposure controls incorporated to minimize and manage potential risk.

Nanotechnology is an example of a fast-paced technology that poses challenges to agencies' policy development and foresight efforts. We have conducted past work looking at the challenges of exercising foresight when addressing potentially significant but somewhat uncertain trends,[5] including technology-based trends that proceed at a high "clockspeed," that is, a (1) faster pace than trends an agency has dealt with previously or (2) quantitative rate of change that is either exponential or exhibits a pattern of doubling or tripling within 3 or 4 years, possibly on a repeated basis.[6] As our prior work has noted, when an agency responsible for ensuring safety faces a set of potentially significant high-clockspeed technology-based trends, it may successfully exercise foresight by carrying out activities such as

- considering what is known about the safety impact of the trend and deciding how to respond to it;
- reducing uncertainty as needed by developing additional evidence about the safety of the trend; and

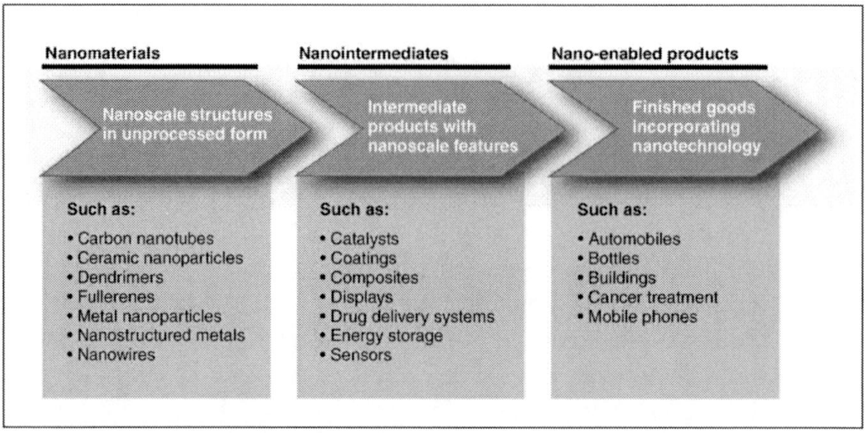

Source: Adapted by GAO from materials produced by Lux Research.

Figure 1. Examples of Nanomaterials as Raw Materials, Intermediates, and Finished Products

- communicating with Congress and others about the trends, agency responses, and policy implications.

Similarly, our 21st Century Challenges report raised concern about whether federal agencies are poised to address fast-paced technology-based challenges.[7] Other foresight literature illustrates the potential future consequences of falling behind a damaging trend that could be countered by early action. These analyses suggest that unless agencies and Congress can stay abreast of technological changes, such as nanotechnology, they may find themselves "in a constant catch-up position and lose the capacity to shape outcomes."[8]

Nanomaterials Currently Enhance Products across a Number of Industry Sectors, and New Uses Continue to Be Developed

Industries around the world are harnessing the properties of nanomaterials for a variety of products across a number of sectors and are expected to continue to find new uses for these materials. Nanomaterials can enter the marketplace as materials themselves, as intermediates that either have nanoscale features or incorporate nanomaterials, and as final nano-enabled products (see figure 1). For example, a manufacturer of clay nanoparticles can provide them to a plastic manufacturer, who can use them to enhance a composite material (an intermediate). The plastic manufacturer can then sell the composite material to an automobile manufacturer, who can e the material to mold parts for cars (nano-enabled products).

As the uses of nanomaterials continue to evolve, the overall market for them is growing, along with the degree to which they are permeating our everyday lives. In 2009, the Woodrow Wilson International Center for Scholars' Project on Emerging Nanotechnologies (Wilson Center) identified a list of more than 1,000 nano-enabled products currently on the market, reflecting a 379 percent increase since this list was first compiled in 2006.[9] The list contains information on products from over 20 countries that can be purchased and used by consumers and provides a baseline for understanding the extent to which nanotechnology is being used. As the Wilson Center has reported, the trend of an increased number of products and applications featuring nanomaterials is also reflected in the number of nanotechnology patents issued by the U.S. Patent and Trademark Office, growing from 125 in 1985 to 4,995 in 2005, which represents a compound annual growth rate of 20 percent. The following is a list of selected industry sectors and some examples of current and potential uses of nanomaterials within each sector that illustrate the ubiquitous nature of these materials in commerce. Because assembling a complete catalog of uses would be difficult in an evolving, dynamic industry, the list is not comprehensive, the examples chosen are simply illustrative, and we have not verified the claims made by the manufacturers of the products used in these examples.

Automotive

From car bodies to exterior coatings to engines on the market today, cars contain numerous enhancements made possible by nanomaterials. In the current marketplace, some bumpers and other auto parts incorporate composite materials containing nanomaterials, such as nanoscale clays, metals, and carbon nanotubes to make these parts stronger, and more fire resistant.[10] Many nano-enabled products in the automotive sector involve the addition of nanoscale ceramic and metal particles to a wide variety of coatings. These nanomaterials provide advantages for coatings over conventional materials, such as the ability to block ultraviolet (UV) light or promote self-cleaning without altering the transparency of the coatings. For example, coatings containing nanoparticles are currently dispersed in paints and pigments to make surfaces stronger, smoother, more scratch and stain resistant, waterproof, or some combination of these and other properties. In addition, carbon nanotubes offer an especially high tensile strength—the ability to withstand a stretching force without breaking—of about 100 times greater than that of steel at one-sixth the weight, and their electrical conductivity can be precisely controlled, which helps prevent the build-up of static electricity. As a result, when a manufacturer of fuel lines adds carbon nanotubes to traditional engineering materials, it results in stronger, safer fuel lines.

In the future, nanomaterials could be used to improve the performance of cars, including reducing wear on engine parts and increasing battery power and fuel efficiency. For example, lubricants that contain certain nanomaterials could provide smaller, stronger, and more stable alternatives to oil-based lubricants. In addition, electrodes—electrical conductors that contain movable electric charges—manufactured at the nanoscale could enable higher-performance rechargeable batteries. For example, according to documents we reviewed, one company that is developing a new lithium-ion battery for electric vehicles uses nanoscale metal oxide materials to create crystallized nanoparticles that may enable this nano-enabled battery to deliver 20 percent more power. Moreover, fuel additives with nanoparticles of cerium oxide could increase diesel engine fuel efficiency.[11] One British company has developed such an application for a fuel-based additive that, due to the size-based properties of cerium nanoparticles, creates a greater surface area for catalyzing the combustion reactions between diesel and air.[12] According to this company, the result is a cleaner burn that converts more fuel to carbon dioxide, produces less noxious exhaust, and deposits less carbon on the engine cylinder walls than other fuel additives. Figure 2 shows examples of some current and potential nanotechnology innovations that may be used in automobiles.

Defense and Aerospace

Nanomaterials are beginning to be used in aerospace applications by manufacturers seeking to take advantage of the electrical and mechanical strength advantages they offer and by the Department of Defense, which is seeking ways to enhance the tools available to its soldiers and the effectiveness of its weapons systems. Nanomaterial polymers are currently being used as sensors that detect very small traces of explosives, which indicate the presence of buried landmines, according to Department officials. In addition, according to documents we reviewed, stronger and lighter planes that are better protected against lightning and fire have been made possible by using carbon nanotubes and other nanostructured materials. For example, one company has created a nanolaminated material used for planes that is comprised of layers of metal alloys that are stronger, lighter, and more energy absorbent than steel. In addition, polymers with embedded silver nanoparticles are helping to keep surfaces, including the interiors of aircraft, free of microbes.[13] The polymers contain nanoscale silver particles that, when added to a product's surface, release ions that kill bacteria existing on the surface.[14] Companies are also introducing nanostructured alternatives to standard copper wiring. For example, one company has developed a process to create highly conductive sheets of fabric and lengths of yarn containing carbon nanotubes that can be used to create wiring and cables for airplanes and satellites that weigh much less than traditional copper wire.

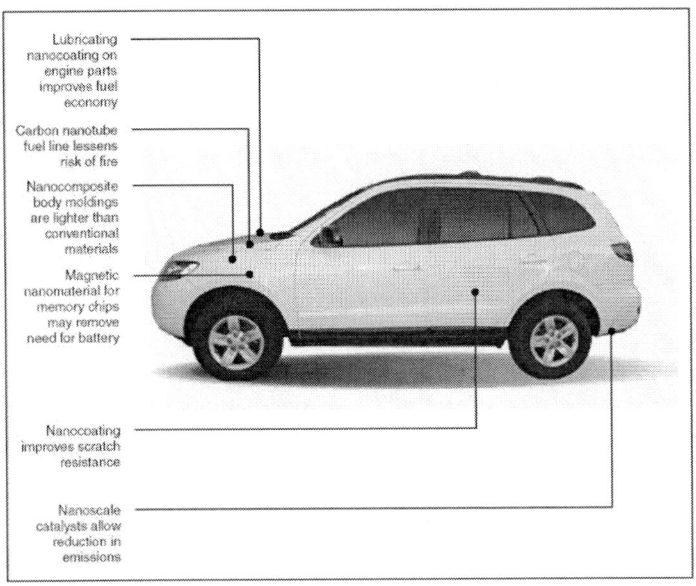

Source: Adapted by GAO from materials produced by Lux Research.

Note: The photo is illustrative and not intended to imply that this particular vehicle currently utilizes the nanotechnology innovations depicted or will in the future.

Figure 2. Examples of Current and Potential Nanotechnology Innovations that May Be Used in an Automobile

In the future, nanomaterials may help enable the development of new applications and products across a wide spectrum in the defense arena, including surveillance devices, explosives and propellants, and uniforms. For example, according to Department of Defense officials and documents we reviewed, nearly "invisible" surveillance may be possible through the incorporation and integration of different nanotechnologies, including radio frequency identification chips; integrated circuits; minute biosensors; and "intelligent" fabrics, films, and surfaces. Miniaturized surveillance techniques under research include using live insects ("spy" bees) tagged with nanomaterials or tiny winged robots that emulate insects to fly into an enemy situation to record data. In addition, more powerful conventional explosives and faster moving missiles may be possible due to the greater amounts of energy provided by nanostructured aluminum. In combination with metal oxides, such as iron oxide, nanostructured aluminum allows many more chemical reactions to occur in a given surface area, increasing the explosive force. Also, nanomaterials such as carbon nanotubes embedded in fabric could allow for lighter uniforms and multifunctional combat suits for soldiers. The uniforms could potentially, for example, change color to match the environment, become rigid casts to protect injuries, or help block bullets and chemical/biological agents. The material could even incorporate sensors that monitor a soldier's condition, or function as drug dispensers activated automatically via radio waves by a remote doctor.

Electronics and Computers

Computers and consumer electronics have also begun to benefit from the advantages nanomaterials offer, including improved display screens and improved electrical conductivity. Carbon nanotubes, quantum dots, [15] and nanoscale layers of polymers can improve the properties of displays. For example, one company has developed an ultra-thin, layered system of polymers that, unlike conventional liquid crystal displays, requires no backlights or filters. The images are brighter and clearer, and the technology could make possible fully bendable plastic displays, according to the company. In addition, since nanomaterials often enhance electrical conductivity, metallic nanoparticles and carbon nanotubes are being used in a growing number of conductive coatings, such as those used for touchscreens and solar cells. According to documents we reviewed, one company sells a transparent conductive coating and a coated film, both incorporating nanowires, which conduct electricity better than traditional materials. The coating and film could eventually replace rare and expensive indium tin oxide, currently the most widely used transparent conductor in the display industry. Moreover, nanomaterials such as lead-free, conductive adhesives could eliminate several steps in manufacturing electronics and could lead eventually to elimination of some or all of the 3,900 tons of toxic, leaded solder used every year by the U.S. electronics industry, according to an EPA document.

In the future, computers and electronic devices could employ nanomaterials to create more efficient data storage and longer-lasting, rechargeable batteries. Memory storage devices could become more powerful through a variety of nanotechnology applications. New methods of storing information electronically are emerging from a variety of applications aimed at increasing the amount of information that can be stored on a given physical space. For example, one company has demonstrated the potential to create high-density memory devices with an estimated storage capacity of 1 terabyte per square inch—more than 200 times higher than the storage density of a DVD—by storing information mechanically using nanoscale probes to punch nanoscale indentations into a thin plastic film.[16] In addition, companies, research institutions, and government labs are working to develop nano-based technology that could perfect "microbatteries," which are smaller, cheaper, and more powerful than batteries currently in use. The greater surface area of the nanowires used in these batteries lowers the internal resistance of the battery and therefore allows greater current flow. Figure 3 shows some examples of current and potential nanotechnology innovations that may be used in a mobile phone.

Energy and Environment

Companies are beginning to use nanomaterials to clean up waste, substitute nonrenewable resources with renewable ones, reduce pollution, and increase the efficiency of solar power. Because nanoscale particles can be more chemically reactive than conventionally scaled particles of the same substance due to their large surface area to volume ratio, these materials can be useful for environmental remediation. Specifically, the increased surface area of various types of ceramic or metal nanomaterials can result in the rapid reduction of contaminant concentrations in soil, water, and air, as pollutants or toxins in these media react with the nanomaterials. Similarly, nanoscale iron is being deployed in a growing number of environmental remediation projects with results that are proving successful so far, according to EPA officials. For example, at one remediation project, researchers injected carbon infused with nanoparticles of iron into contaminated soil and found that the nanoparticles made the resulting material more effective at absorbing contaminants than similar materials without the nanoparticles. In addition, nanomaterials are being used to create packaging materials made from waste. For example, one company produces nanoparticle paper coatings made from renewable natural starches that can replace conventional material in paper coatings, which is typically made from nonrenewable petroleum. Nanomaterials are also being used to improve automotive catalytic converters, which feature nano-enabled catalysts that reduce air pollution more efficiently. One company is manufacturing a catalyst consisting of nanostructures with surface areas much higher than traditional materials and that allows catalytic converters to remain effective under prolonged exposure to high temperatures, resulting in more stable, durable, and cost-effective products. In the energy arena, nano-enabled thin-film and photovoltaic technologies are making solar power more efficient. For example, one company has reported gains in the ability of its thin-film solar cell materials to absorb light, because the structure of the nanomaterial is much smaller than the wavelength of light, which allows it to act like an antenna that concentrates, absorbs, and transfers energy with high efficiency.

In the future, nanomaterials could help deliver alternative forms of energy, cleaner water, and more efficient energy transmission. Using nanoscale catalysts, hydrogen—an alternative form of energy—could be produced from water more efficiently. For example, a company has developed a photoelectrode that uses nanoscale material and converts sunlight into hydrogen six times more efficiently than its conventionally scaled equivalent.[17] In addition, nanotechnology-enabled water desalination and filtration systems may offer affordable, scalable, and portable water filtration in the future. Filters, comprised of nanoscale pores which incorporate a wide variety of nanomaterials—including nanoparticles made of aluminum oxide, iron, and gold, and carbon nanotubes—have the potential to allow water molecules to pass through, but screen out larger molecules, such as salt ions and other impurities such as bacteria, viruses, heavy metals, and organic material. In addition, nanoparticles could be used to improve the efficiency of energy transmission by increasing the capacity and durability of insulation for underground electrical cables, allowing cables of smaller diameter to carry the same power as larger cables and to last longer. For example, one company's research shows that cable insulation treated with nanocomposites containing nanosilica have about 100 times longer voltage endurance compared to untreated material. In addition, researchers have demonstrated that carbon nanotube fiber bundles could carry 100 times more electrical current than the leading transmission wires, without as much energy loss. Moreover, one study predicts that if energy transmission losses could be reduced from the current 7 percent using copper wires to 6 percent by using carbon nanotube fibers, the annual energy savings in the United States would be equal to 24 million barrels of oil.

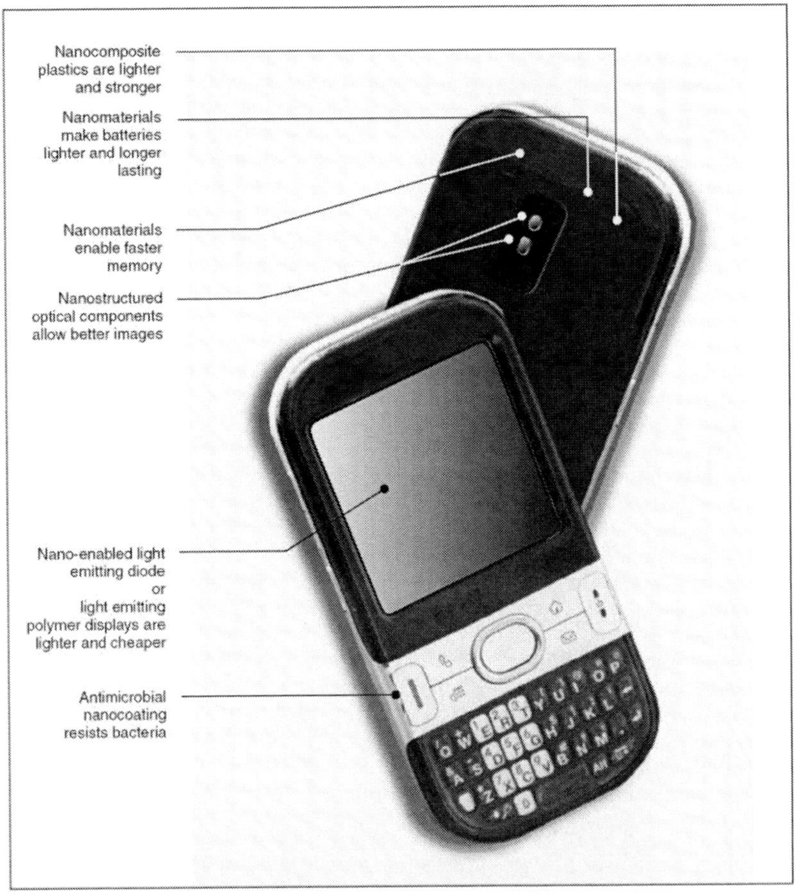

Source: Adapted by GAO from materials produced by Lux Research.

Note: The photo is illustrative and not intended to imply that this particular phone currently utilizes the nanotechnology innovations depicted or will in the future.

Figure 3. Examples of Current and Potential Nanotechnology Innovations That May Be Used in a Mobile Phone

Food and Agriculture

Nanomaterials are currently appearing in food packaging and food supplements.[18] Specifically, nanomaterials are being used in food packaging, where applications such as antimicrobial nanofilms—thin layers of substances meant to hamper the growth of bacteria and fungi—are intended to bolster food safety. Also, composite materials made of nanoclays embedded in nylon can offer strong oxygen and carbon dioxide barriers and have been used in plastic bottles and films for packaging food and beverages. For example, one company produces a nylon and clay nanocomposite used as a flexible, puncture-resistant oxygen barrier for beer and carbonated beverage bottles; in packaging for processed meats and cheeses; and in coatings for paper packaging for juice or dairy products. Moreover, products such as cutting boards and food containers have been infused with nanosilver—which is known for its antimicrobial properties. In addition, encapsulation—the process of using one material to deliver another material inside the human body—has been in use for decades but is being improved with nanomaterials. Nanoencapsulated food products and supplements can target nutrients, release drugs on a controlled schedule, and mask tastes. For example, some vitamins can be difficult to deliver in beverages because they degrade and may not be easily absorbed by the body. One company has developed nanoscale structures to deliver the vitamin to the digestive system, making it easier for absorption to occur. Another manufacturer has used nanocapsules to incorporate certain fatty acids that have purported health benefits into bread. The company claims the acids in the nanocapsules bypass the taste buds, emerging only after the nanocapsules reach the stomach, thus avoiding any unpleasant taste.

In the future, manufactured nanomaterials could be used to enhance agriculture; monitor food quality and freshness; improve the ability to track food products from point of origin to retail sale; and modify the taste, texture, and fat content of food. Nanomaterials are being developed to more efficiently and safely administer pesticides, herbicides, and fertilizers by controlling more precisely when and where they are released. In addition, researchers are developing a nanoscale powder that can retain water better than other materials and allows fertilizers to gradually release nutrients for crops or grass, according to the Wilson Center. In addition, researchers have developed nanobiosensors using nanoscale particles for detecting bacteria, such as salmonella, in water and liquid food. Their work could lead to nanosensors that could be used in fields to monitor for bacterial contamination of crops, such as spinach, lettuce, and tomatoes, potentially reducing the spread of foodborne illnesses. In addition, electrically conductive inks containing nanomaterials could be used to print radio-frequency identification tags, which could be integrated into packaging for food products, potentially resulting in improved food security and better inventory tracking and management. Figure 4 shows some examples of current and potential nanotechnology innovations that may be used in a drink bottle.

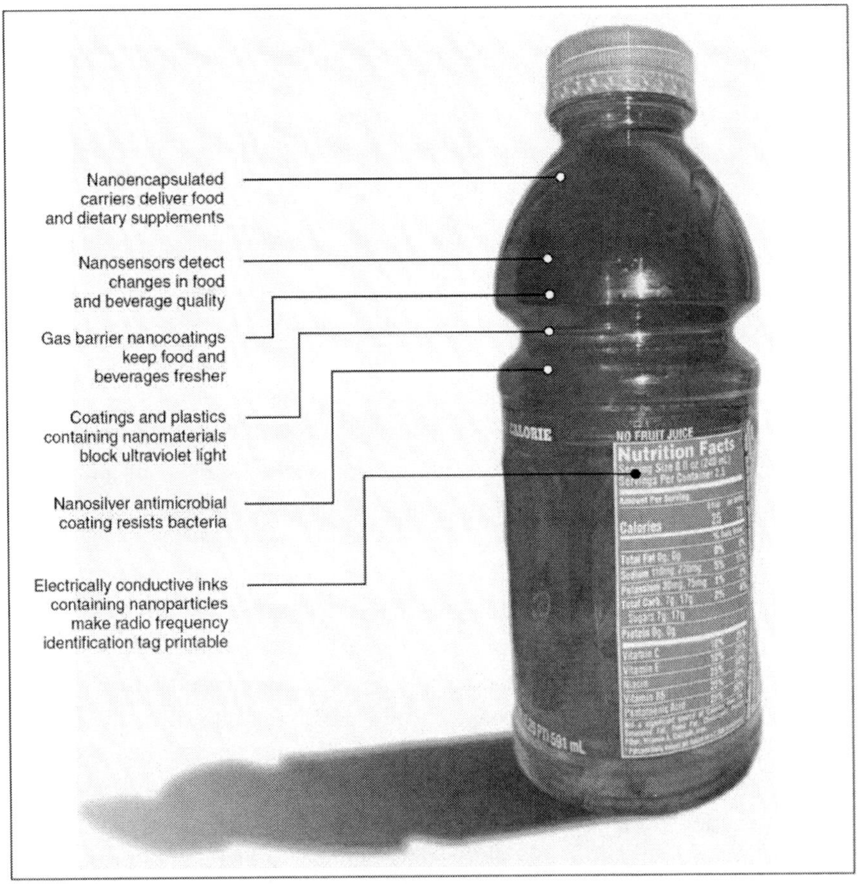

Source: Adapted by GAO from materials produced by Lux Research.

Note: The photo is illustrative and not intended to imply that this particular juice bottle currently utilizes the nanotechnology innovations depicted or will in the future.

Figure 4. Examples of Current and Potential Nanotechnology Innovations That May Be Used in a Drink Bottle

Housing and Construction

Materials and coatings are currently making buildings and homes cleaner and stronger, and in the future will allow them to operate with higher energy efficiency, according to documents we reviewed. Protective coatings and materials that incorporate nanoparticles of titanium dioxide are being used to manage heat and light by blocking UV light from the sun's rays and are taking on self-cleaning properties through a photocatalytic effect.[19] For example, titanium dioxide is being added to paints, cements, windows, tiles, and other products for its sterilizing and deodorizing properties. Additionally, as titanium dioxide is exposed to UV light, it becomes increasingly hydrophilic—attractive to water—and is therefore being used for antifogging coatings or self-cleaning windows. Nanomaterials are also proving beneficial to the construction industry by, for example, making steel tougher and concrete stronger, more durable, and more easily placed. For example, one company has created a structural material with a grain size reduced to the 100 nanometer scale, which it claims has a strength-to-density ratio four times that of the toughest titanium alloys and also resists corrosion. Inside the walls of buildings, insulation made from nanomaterials is providing high thermal performance at minimal weight and thickness. In addition, nanomaterials are being incorporated into some air monitoring technologies, air purification products, and energy-efficient air conditioning systems for residential, commercial, and industrial settings. For example, some air filters that are on the market use nanomaterials to clean air better than conventional materials.

In the future, nanoparticle coatings on windows and buildings could retain energy from the sun for later release. For example, researchers working on phase change materials—materials which absorb and release thermal energy—have found that when graphite nanofibers are blended into these materials the nanofibers improve the material's thermal performance. The result could be cheaper and more efficient uses of these materials for solar energy storage. In addition, nanomaterials may offer approaches that enable materials to "self-heal" by incorporating, for example, nanocontainers of a repair substance (e.g., an epoxy) throughout the material. When a crack or corrosion reaches a nanocontainer, it could be designed to open and release its repair material to fill the gap and seal the crack. Figure 5 shows some examples of current and potential nanotechnology innovations that may be used in a house.

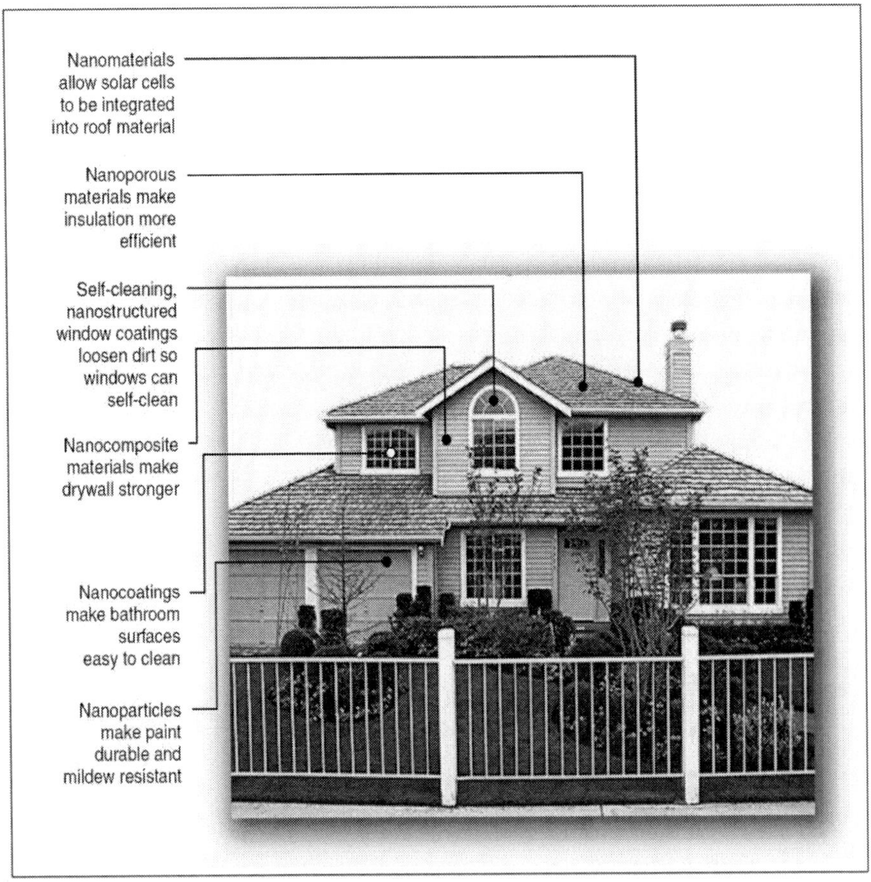

Source: Adapted by GAO from materials produced by Lux Research.

Note: The photo is illustrative and not intended to imply that this particular house currently utilizes the nanotechnology innovations depicted or will in the future.

Figure 5. Examples of Current and Potential Nanotechnology Innovations That May Be Used in a House

Medical and Pharmaceutical Nanotechnology is important to the medical and pharmaceutical industry because the extremely small size of nanomaterials makes possible medical interventions that can be directed to individual cell types, allowing for better diagnosis, treatment, and prevention of cancer and other deadly diseases.[20] Current disease detection efforts include the use of nanoscale sensors to identify biomarkers, such as altered genes, that may provide an early indicator of cancer. Doctors are also using nanomaterials as markers to enhance images from deep inside human tissue, allowing them to track particles to the site of a tumor, resulting in earlier detection of tumors. Certain nanomaterials such as polymer nanoparticles are being used to treat cancer by delivering medication directly to tumors while sparing healthy tissue. In addition, silver nanocrystals are being used in antimicrobial wound dressings, thereby requiring fewer dressing changes and causing patients less pain.

In the future, nanomaterials could be used to help doctors better diagnose and treat disease. In diagnosis, nanomaterials hold promise for showing the presence, location, and contours of cardiovascular and neurological disease, and small tumors. For example, researchers could use metallic and magnetic nanoparticles to enhance imaging, the results of which can be used to guide surgical procedures and to monitor the effectiveness of nonsurgical therapies in reversing the disease or slowing its progression. In the future, sensors implanted or delivered with a drug could allow for continuous and detailed health monitoring so disease might be managed better, turning a drug into a multifunctional tool for diagnosis and treatment. For example, bio-sensors could be attached to targeted drugs and linked to a mechanism that reports the body's condition. Furthermore, according to the National Institutes of Health, gold nanoshells are being developed to simultaneously image and destroy cancer cells using infrared light. Nanoshells can be designed to absorb light of different frequencies, generating heat. Once the cancer cells take up the nanoshells, scientists apply near-infrared light that is absorbed by the nanoshells, creating an intense heat inside the tumor that selectively kills tumor cells without disturbing neighboring healthy cells. Such a targeted delivery approach could reduce the amount of chemotherapy drug needed to kill cancer cells, potentially reducing the side effects of chemotherapy. Medical researchers are also exploring the use of nanomaterials to deliver molecules and growth factors to promote better healing for burns and wounds that heal without scars. For example, Department of Defense researchers have conducted tests in animals using nanofiber mesh scaffolds to treat bone, nerve, cartilage, and muscle injuries and have reported that preclinical data from the studies indicate improved healing. Other nanofibers are being developed for medical use as mesh barriers to stop the flow of blood and other fluids more quickly and effectively.

Personal Care, Cosmetics, and Other Consumer Products

Nanomaterials are currently being used in a variety of personal care items, cosmetics, and other consumer products.[21] These products include sunscreens that contain nanoscale titanium dioxides and zinc oxides, which act as physical filters that absorb UV light. Because these nanomaterials are smaller than the wavelength of light, they make sunscreens transparent instead of opaque, and they may also adhere better when applied and absorb harmful ultraviolet rays more effectively than conventional sunscreens, according to stakeholders and documents we reviewed. In addition, nanomaterials are being incorporated into cosmetics, such as an anti-aging cream, which allows the active ingredients to penetrate deep into the skin where they can be most effectively administered, according to the manufacturer. Nanomaterials are also being used in a wide range of other consumer products. For example, companies are using carbon nanotubes to reinforce a variety of sporting goods, such as bicycle frames, tennis rackets, baseball bats, and hockey sticks, because they offer greater strength and reduced weight, while retaining, or even increasing, stiffness. Companies are using other nanomaterials to improve the performance of products such as ski wax and tennis balls. For example, a nanomaterial coating decreases the gas permeability in tennis balls and therefore allows the balls to maintain pressure for longer periods of time, according to the company producing the coating. Nanomaterials are also being used in coatings to make fabrics and clothing stain and water resistant. For example, one company embeds nanomaterials on the surface of fabric fibers, creating a cushion of air around them. The fabric allows sweat to pass out, while also causing surface water to bead up and roll off. Another company has developed socks treated with nanosilver for its antimicrobial properties.

In the future, consumers may benefit from advanced applications that could emerge from nanomaterial research occurring in a variety of sectors. For example, developments in the health arena could lead to new, beneficial pharmaceutical therapies designed to treat aging and age-related disease. In addition, according to documents we reviewed, researchers are working to make textiles functional by combining manufactured nanomaterials with materials that react to light to create power-generating clothing and nanosilver could be used in textiles to treat skin conditions. Researchers are also developing nano-enabled textile surfaces that can remove scratches and scuff marks, as well as decolorize red wine spills.

POTENTIAL RISKS TO HUMAN HEALTH AND THE ENVIRONMENT FROM NANOMATERIALS DEPEND ON TOXICITY AND EXPOSURE, AND CURRENT UNDERSTANDING OF THE RISKS IS LIMITED

The properties of nanomaterials affect their toxicity and, in turn, their risks to human health and the environment. Furthermore, the risk of nanomaterials also depends on the extent and route of exposure to nanomaterials, but current understanding of nanomaterial toxicity and exposure is limited, according to the studies we reviewed.

The Toxicity of Individual Nanomaterials May Vary According to Their Properties and Affects Their Risks

The toxicity of each nanomaterial may vary according to a combination of the individual properties of these materials—including size, shape, surface area, and ability to react with other chemicals—and these properties affect the potential risks posed by nanomaterials, according to some of the studies we reviewed. The properties of a nanomaterial may differ from the properties of conventionally scaled material of the same composition. For example, the properties of conventionally scaled gold have been well characterized: gold is metallic yellow in color and does not readily react with other chemicals. As a nanoparticle, however, gold can vary in color from red to black and become highly reactive. The following are examples of how toxicity may be affected by the properties of nanomaterials as compared with their conventionally scaled counterparts:

- *Size.* Research assessing the role of particle size on toxicity has generally found that some nanoscale (<100 nanometers) particles are more toxic and can cause more inflammation than conventionally scaled particles of the same composition. Specifically, some research indicates that the toxicity of certain nanomaterials, such as some forms of carbon nanotubes and nanoscale titanium dioxide, may pose a risk to human health because these materials, as a result of their small size, may be able to penetrate cell walls, causing cell inflammation and potentially leading to certain diseases. For example, the small size of these nanomaterials may allow them to penetrate deeper into lung tissue, potentially causing more damage, according to some of the studies we reviewed. In addition, some nanomaterials may disperse differently into the environment than conventionally scaled materials of the same composition because of their size. However, according to EPA, the small particle size may also cause the nanomaterials to agglomerate, which may make it more difficult for them to penetrate deep lung tissue.

- *Shape.* Nanomaterials may be produced in a wide variety of shapes, including spheres, tubes, threads, and sheets, as well as more ornate forms, such as dumb-bells. The shape of nanomaterials may be connected to the type of health risks they may pose. For example, some carbon nanotubes resemble asbestos fibers. When inhaled by people, asbestos fibers are known to cause mesothelioma—which is a disease associated with asbestos exposure. The similarity of these carbon nanotubes to asbestos fibers has caused researchers to question if exposure to such nanomaterials may lead to a similar disease. Furthermore, a study has shown that exposing the abdominal cavity of mice to certain long carbon nanotubes may be linked with inflammation of the abdominal wall. The abdominal cavity in mice is often used as a surrogate for understanding how the mesothelial lining of the human chest cavity will react to substances.
- *Surface area and reactivity.* Nanomaterials may also be more reactive with other chemicals than similar conventionally scaled materials because nanomaterials have a higher surface area-to-mass ratio, providing more area by weight for chemical reactions to occur. Some studies have found that because of this increased reactivity, some nanoscale particles may be potentially explosive and/or photoactive—that is, sunlight triggers a chemical reaction in them. For example, some nanomaterials—such as nanoscale titanium dioxide and silicon dioxide—may explode if finely dispersed in the air and they come into contact with a sufficiently strong ignition source. However, in general, the extent to which such nanoscale dusts may be more explosive than larger size dusts of the same composition is not fully known, according to the National Institute for Occupational Safety and Health. Other research has shown that particle surface area is a better predictor of toxic response to inhaled particles than is particle mass. For example, research into nanoscale titanium dioxide in mice and rats has shown that particle surface area seems to be a more

appropriate measure for comparing the effects of different-sized particles, provided they are of the same chemical structure.

Risk of Nanomaterials Is Also Affected by the Route and Extent of Exposure

In addition to toxicity, the risk that nanomaterials pose to humans and the environment is also affected by the route and extent of exposure to such materials. Nanomaterials can enter the human body through three primary routes: inhalation, ingestion, and dermal penetration.[22]

- Inhalation is the most common route of exposure to airborne nanoparticles, according to the National Institute of Occupational Health and Safety. For example, workers may inhale nanomaterials while producing them if the appropriate safety devices are not used, while consumers may inhale nanomaterials when using products containing nanomaterials, such as spray versions of sunscreens containing nanoscale titanium dioxide. According to officials at the National Institutes of Health, although the vast majority of inhaled particles enter the pulmonary tract, evidence from studies on laboratory animals suggest that some inhaled nanomaterials may travel via the nasal nerves to the brain and gain access to the blood, nervous system, and other organs, according to studies we reviewed.
- Ingestion of nanomaterials may occur from unintentional hand-to-mouth transfer of nanomaterials or from the intentional ingestion of nanomaterials.[23] Ingestion may also accompany inhalation exposure because particles that are cleared from the respiratory tract can be swallowed. A large fraction of nanoparticles, after ingestion, rapidly pass out of the body; however, according to some of the studies we reviewed, a small amount may be taken up by the body and then migrate into organs. The effect of these small amounts of ingested nanomaterials is currently unknown, but concerns have arisen from a growing body of evidence which indicates that certain types of nanoparticles may cross cellular barriers.
- Nanomaterials may also be absorbed through the skin. For example, one laboratory study has shown that certain nanomaterials have penetrated layers of pig skin within 24 hours of exposure. In addition, some cosmetics and sunscreens—among the first commercial

> products to incorporate nanomaterials—contain nanoscale titanium dioxide to increase the ultraviolet light-blocking power of the product. The nano titanium dioxide is believed to be less toxic than other chemicals that have been used to provide ultraviolet protection in sunscreens. However, according to some of the studies we reviewed, concerns have been raised that nanomaterials in sunscreens could penetrate damaged skin. In contrast, according to officials at the National Institutes of Health, there are several studies that have found little dermal penetration from nanomaterials when applied to undamaged skin. According to some stakeholders we spoke to, given the known hazards of sun exposure, sunscreens containing nanomaterials may be reasonable choices for the protection that they provide to consumers from sun exposure.

In addition to the route of exposure, the extent of exposure—that is the frequency and magnitude—to consumers and workers also affects the risks posed by nanomaterials. Workers may be accidentally exposed to nanomaterials during the production of nanomaterials or products containing them, as well as during use, disposal or recycling of these products. At present, there is insufficient information on the number of workers exposed to nanomaterials in the work place or the effects on human health of such exposure, according to the European Agency for Safety and Health at Work. In addition, because nanomaterials have applications in many consumer products and the use of such materials in products is increasing, consumers have an increasing chance of exposure to these materials. For example, consumers may now purchase appliances such as washing machines coated with silver nanomaterials purported to kill bacteria. When consumers purchase such a machine, their clothing will be exposed to the silver nanomaterials, thus increasing their exposure to nanomaterials. Similarly, consumers may now purchase socks containing nanosilver, which exposes them to this nanomaterial. According to EPA officials, occupational exposure is a particular concern and warrants attention because the exposure and risk to workers is potentially greater than the risk to consumers.[24]

In addition to humans, the environment may also be exposed to nanomaterials through releases into the water, air, and soil, during the manufacture, use, or disposal of these materials. For example, nanomaterials could enter water through discharges from production facilities. In addition, when nanomaterials are used in pharmaceuticals, cosmetics, and sunscreens, the nanomaterials could enter water via the sewage system during washing, showering, or swimming after having been applied to the skin and may eventually end up in a waste water treatment plant. These nanomaterials, if antibacterial in nature and if released in sufficient amounts, could potentially interfere with beneficial bacteria in sewage and waste water treatment plants and could also contaminate water intended for re-use, according to some of the studies that we reviewed. Moreover, some researchers have raised serious concerns that antibacterial nanomaterials will pose toxicity risks to human health and to environmental systems into which waste products are released. In addition, according to research, unused cosmetics are most likely to be disposed of in household waste, which may be incinerated, potentially putting nanomaterials into the air, or put in a landfill, potentially leaching out of the landfill into the water. In addition, nanomaterials that are currently being used to treat polluted water will result in releases of the materials into water and soil. For example, iron nanoparticles are being used to treat polluted water. According to EPA officials, although little is known about how these particles move through the environment, they are expected to react with contaminants or with naturally occurring substances in water and become iron oxides. Figure 6 shows the potential exposures to humans and the environment throughout the lifecycle of nanomaterials.

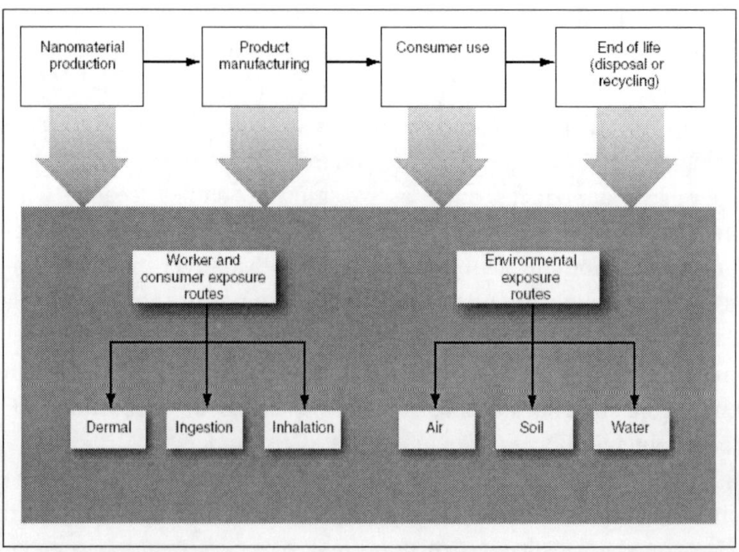

Source: Adapted by GAO from materials produced for the European Parliament's Committee on the Environment, Public Health and Food Safety.

Figure 6. Potential Exposure Routes throughout the Life Cycle of Nanomaterials

Currently, it is difficult to assess the risk of nanomaterials that are released into the environment because these materials are so varied and it is difficult to make generalizations about how they will behave once they are released, according to EPA officials. Specifically, it is unclear whether the nanomaterials will (1) stay suspended, (2) aggregate or cluster together to form larger particles, (3) dissolve or further break down, or (4) react with natural materials found in the environment. For example, the release of carbon nanotubes, nanoparticles of iron and titanium dioxide, or fullerenes—which are nanoscale spheres of carbon—into water may result in their aggregation, according to some of the studies we reviewed. These larger aggregates may have different toxicological properties when compared to those exhibited by the original nanomaterials. The risk posed by some nanomaterials is presumed to decrease if they aggregate because the nanomaterials may grow to the size of conventionally scaled substances, according to some of the studies we reviewed. However, the extent of aggregation may be limited because many nanomaterials receive coatings to decrease the aggregation of these materials. In addition, some nanomaterials may react with the environment and eventually build up in the environment, according to some of the studies we reviewed. Specifically, some nanomaterials may become attached to and continue to build up in the soil, depending on the nanomaterial characteristics and the characteristics of the soil. Some nanomaterials may also bioaccumulate in organisms, according to EPA.

Understanding of the Risks Posed by Nanomaterials Is Limited by Several Factors

Current understanding of the risks that nanomaterials may pose is limited by several factors, including the limited amount of research that has been conducted to date and a lack of tools and methods needed to conduct additional research. As a result, predicting and assessing the potential hazards, exposures, and resulting risks from nanomaterials is difficult. Although the number of studies that have focused on assessing the risks of nanomaterials has increased over the past 5 years (see figure 7), the studies completed to date have yielded limited risk information, according to EPA officials and other stakeholders that we spoke with, and our review of these studies. Some of these limitations include the following:

- The findings from completed toxicity studies of a nanomaterial constructed in one manner may not be applicable to understanding the risks posed by the same nanomaterial constructed in a different manner and, therefore, studies of similar nanomaterials may not be comparable. For example, carbon nanotubes may be produced in several ways, each with its own potential level of toxicity so that the results of a study for one type of carbon nanotube may not be comparable to the results of a study of a different type of carbon nanotube. Similarly, some early studies of carbon nanotubes did not specify the length of the nanotubes being studied, making it difficult to compare the results of those studies with subsequent carbon nanotube studies, according to stakeholders. This is important because researchers now know that different nanotube lengths may pose different risks.
- The studies that have been conducted have focused more extensively on some nanomaterials than others. For example, certain silica nanoparticles and carbon black are among the best studied nanomaterials, according to EPA. In contrast, less is known about nanomaterials such as nanoscale aluminum oxide and nanoclays. Therefore, little or no information is known about the risks of these types of nanomaterials.

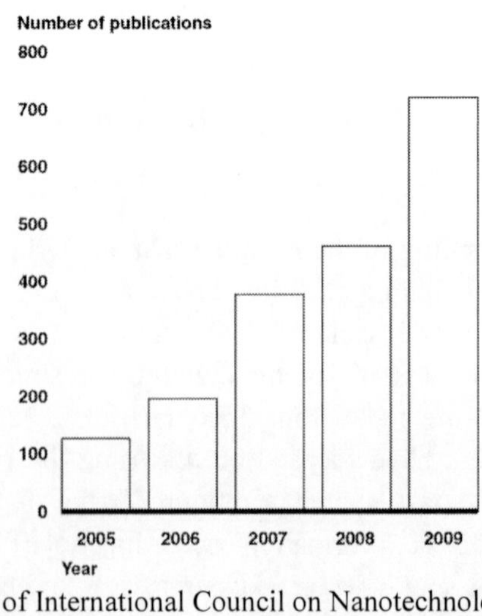

Source: GAO analysis of International Council on Nanotechnology data.

Figure 7. The Increase in Environment and Human Safety Research Relating to Nanomaterials since 2005

Additional efforts to study the risks from nanomaterials will also be hampered because certain tools necessary to conduct these studies are lacking. Specifically, according to studies we reviewed, research on nanomaterials depends on the availability of tools, such as models or measurement technologies, to characterize or describe the nanomaterials' main qualities. However, although some tools are available, the scientific community does not currently possess all the needed tools to do so, and it will require extensive research to develop these tools. Additionally, lack of data and appropriate models also limits our ability to study the risks posed by nanomaterials, according to some of the studies we reviewed and stakeholders that we spoke with. While researchers have developed models for conventionally scaled chemicals that predict their characteristics based on the characteristics of similar, or analogous, chemicals, no such models exist yet for nanomaterials. For example, as mentioned earlier, free nanoparticles may aggregate in the natural environment, forming larger structures that may have different toxicological properties to those exhibited by the original nanoform, but researchers lack models to accurately predict how, when, and with which nanomaterials this aggregation will occur. Moreover, according to stakeholders we spoke to, small changes in the characteristics of some nanomaterials, such as a 10 percent change in their size, may alter the toxicity of the nanomaterials. The effect of such a small change compounds the difficulty in creating predictive models of nanomaterial toxicity.

EPA Has Taken a Multipronged Approach to Managing the Potential Risks of Nanomaterials but Faces Various Challenges in Regulating These Materials

EPA has taken a variety of actions to better understand and regulate the risks of nanomaterials, including conducting research and asking companies to voluntarily provide information about the nanomaterials that they produce or use. Although EPA has taken some regulatory action under its existing statutory framework with regard to nanomaterials, its authority to do so varies depending on the statute that it is using to regulate specific nanomaterials.[25] Moreover, the agency faces additional technical and informational challenges that may impact its ability to regulate nanomaterials effectively.

EPA Has Ongoing Research Efforts Related to Nanomaterials

In June 2009, EPA issued its Nanomaterial Research Strategy, which lays out the agency's plans for research to understand the potential human health and environmental impacts from exposure to nanomaterials, as well as how certain nanomaterials can be used in environmental protection applications, such as remediating contaminated waste. The strategy builds upon a body of research already conducted by EPA in areas such as ultrafine particulate exposure and toxicity, fate and transport modeling, life cycle assessment, and green chemistry.[26] EPA's strategy states that the agency's research efforts will advance two key objectives: (1) develop approaches for identifying and addressing any hazardous properties, while maintaining beneficial properties, before a nanomaterial enters the environment and (2) identify whether, once a nanomaterial enters the environment, it presents environmental risks. EPA stated that it plans to pursue these objectives from a life cycle perspective—from the production of a nanomaterial, through its use in products, and as it is disposed of or recycled. Ultimately, EPA plans to develop models and other tools to enable it to predict the risks posed by various types of nanomaterials. According to the strategy, EPA's research efforts will be coordinated with those of other federal agencies. For example, EPA's laboratories are collaborating with the National Institutes of Health to conduct research on, among other things, the health effects of carbon nanotubes. According to EPA, its research builds on and is consistent with the scientific needs identified by the NNI's Nanotechnology Environmental and Health Implications working group and in EPA's 2007 Nanotechnology White Paper.

EPA is also coordinating with international organizations, such as the Organisation for Economic Co-operation and Development (OECD) and the International Organization for Standardization (ISO),[27] on nanomaterials research. Specifically, the OECD established the Working Party on Manufactured Nanomaterials in September 2006, with EPA as a member and the initial chair of the working party. This working party is engaged in a variety of projects to further the understanding of the properties and risks of nanoscale materials and how to mitigate exposures and potential risks. For example, one project involves a program for testing the safety of a set of 14 nanomaterials. Specifically, member countries have agreed to develop certain data for a group of 14 nanomaterials selected by the OECD working party, in part, because they are in commerce or close to commercial use.[28] As part of this effort, EPA has the lead for the testing of fullerenes, single-walled carbon nanotubes, multiwalled carbon nanotubes, silver nanoparticles, and nano cerium oxide, among others. In addition, EPA is participating in several ISO working groups for nanomaterials. ISO has established a technical committee to develop international standards for, among other things, nanotechnology terminology, specifications for reference materials, and test methodologies.

Under TSCA, EPA Has Regulated Some Nanomaterials as New Chemicals or New Uses, but Some Nanomaterials May Be Entering the Market without EPA Review

Over the last 3 years, EPA's approach for regulating nanomaterials under TSCA has been evolving as more information has become available on the potential risks. In January 2008, EPA launched a voluntary program called the Nanoscale Material Stewardship Program. Under this program, EPA posted a notice in the *Federal Register* asking manufacturers and processors of nanomaterials to submit existing information on the nanomaterials they produce or use to help EPA better understand the human health and environmental risks from these substances. Thirty-one companies voluntarily provided information on 132 nanomaterials, according to EPA officials. In its interim report on this program, issued in January 2009, EPA noted that although the program provided useful information regarding certain nanomaterials in commerce, a significant number of environmental health and safety data gaps remain. For example, as part of the voluntary program, EPA estimated that companies provided information on only about 10 percent of the nanomaterials that are likely to be commercially available. In addition, EPA reported that its review of data submitted through the program revealed instances in which the details of the manufacturing, processing, and use of the nanomaterials, as well as exposure and toxicity data, were not provided. This further reduced the usefulness of the information received because exposure and toxicity data are two of the major categories of information that EPA had identified as being needed to better inform its risk assessments of nanomaterials. EPA concluded from the low response rate that most companies were not inclined to voluntarily supply information on their nanomaterials.

In January 2008, EPA released a document entitled *TSCA Inventory Status of Nanoscale Substances—General Approach*, which addressed whether nanomaterials constituted new chemicals for the purpose of regulation under TSCA. TSCA provides EPA with different authorities for regulating new chemicals and existing chemicals. New chemicals are those that are not already listed on the TSCA inventory, which is a list of chemical substances manufactured or processed in the United States. Existing chemicals are those already in commerce, including about 62,000 which were already in commerce when EPA began reviewing chemicals in 1979. In general, existing chemicals can be manufactured or processed without any notification to EPA. By contrast, companies intending to manufacture a new chemical must generally submit a notice to EPA before manufacturing or producing the chemical. In its 2008 document, EPA stated that a nanomaterial is a new chemical for purposes of regulation under TSCA only if it does not have the same "molecular identity" as a chemical already on the inventory. Under TSCA, a chemical is defined in terms of its particular molecular identity. Although molecular identity is not defined in the statute, EPA considers chemicals to have different molecular identities when, for example, they represent different allotropes—a variant of a substance consisting of only one type of atom—or isotopes. [29] According to EPA officials, EPA generally does not consider the properties—such as size, shape, and reactivity—of a chemical in establishing its molecular identity. Thus, because titanium dioxide is already listed on the TSCA inventory, nanoscale versions of titanium dioxide, which have the same molecular formula, would not be considered a new chemical under TSCA, despite having a different size or shape, different physical and chemical properties, and potentially different risks. However, fullerenes—a class of nanomaterials made of spheres of carbon—would be considered a new chemical because they represent a different allotrope, or molecular arrangement of carbon atoms, than those chemicals already listed on the inventory.

If EPA makes certain findings, on the basis of information presented in a premanufacture notice, it may control the manufacture, processing, distribution in commerce, use, and disposal of the chemical. The agency sometimes issues a consent order to the company that places conditions on the use of the chemical or requires the company to generate more information on the chemical's health and environmental effects. Since 2005, the agency has received over 90 premanufacture notices for nanomaterials under TSCA, according to EPA officials. EPA officials also told us that about 20 of these notices were requests to be exempt from the full new chemical review process based on regulatory exemptions for substances that met specific low release and exposure criteria or which were produced at low volumes.

TSCA also authorizes EPA to issue rules addressing new uses of certain materials—known as Significant New Use Rules (SNUR). These rules identify new uses of existing chemicals that could affect the nature of human and environmental exposure to the substance. If a company wants to use a chemical in a way that has been designated as a significant new use, it must submit a Significant New Use Notice to EPA. For example, if EPA determines that manufacturing a chemical in a powder form instead of a liquid form would be a significant new use of that chemical, the company planning on manufacturing the chemical in a powder form would have to notify EPA. Upon receipt of a notice, EPA has 90 days to evaluate the intended use and, if warranted, to prohibit or limit it before it occurs. In 2008, EPA issued two such rules for nanomaterials. Specifically, having received premanufacture notices for nanoscale versions of siloxane-modified silica and alumina particles, EPA determined that certain uses of these chemicals, including use without employing personal protective equipment, as a powder, and uses different from those described in the premanufacture notices, were significant new uses.

In 2008, EPA entered into consent orders with a manufacturer of a specific type of carbon nanotubes that placed conditions on the use of that manufacturer's carbon nanotubes. EPA was unable to determine the potential for human health effects of these nanomaterials based on the information available in the premanufacture notices and determined that the uncontrolled manufacture, import, processing, distribution, use, or disposal of these nanomaterials may present an unreasonable risk to human health. Accordingly, EPA imposed exposure and release controls on the manufacture of these nanomaterials in addition to certain testing requirements. Subsequently, in November 2009, EPA proposed SNURs for these nanomaterials, making the limitations articulated in the consent orders applicable to all companies that might seek to manufacture them, and in January extended the comment period until February 2010.[30] As of March 2010, no final rule had been issued, but according to EPA, the agency is in the process of issuing the final SNURs after considering public comment. Until the SNURs are finalized, carbon nanotubes produced by manufacturers other than those bound by the consent orders may be entering the market without EPA review of available information on their potential risk. However, according to EPA, no manufacturer or importer has been able to demonstrate that their carbon nanotubes are chemically identical to another manufacturer's carbon nanotubes; hence the agency has treated all carbon nanotubes as unique chemical substances for the purpose of listing them on the TSCA chemical inventory.

In the fall of 2009, EPA announced it would reconsider the policy described in its January 2008 document, *TSCA Inventory Status of Nanoscale*

Substances—General Approach, and subsequently announced it planned to develop a SNUR to regulate nanoscale versions of conventionally scaled chemicals that are already on the TSCA inventory as a significant new use of that chemical. The agency intends to propose this rule in December 2010. EPA stated the agency would determine the existing uses of nanomaterials by using information submitted through the voluntary Nanoscale Materials Stewardship Program and other sources. EPA officials told us that issuing a SNUR would allow the agency to regulate nano versions of chemicals already on the TSCA inventory the same way it would regulate a new chemical. One problem that EPA may face in issuing such a SNUR is that many uses of nanomaterials are no longer new because nanomaterials are rapidly entering the market, according to stakeholders we spoke with.

TSCA also gives EPA authority to issue rules requiring companies to submit certain information about chemicals. EPA plans to issue one such rule for nanomaterials that would require manufacturers to provide information on production volume, methods of manufacture and processing, and exposure and release, as well as available health and safety studies.[31] Evaluation of this information will provide EPA with an opportunity to consider appropriate action under TSCA to reduce unreasonable risks to human health or the environment, according to EPA. This rule may also help them collect information on nanomaterials not covered by the SNUR discussed above. EPA intends to propose this rule in December 2010. This, however, raises the concern that, in the meantime, nanomaterials may be entering the market without the scrutiny these materials may merit. Furthermore, stakeholders and EPA officials point out that the completeness of information collected under a reporting rule may be limited because the current definition of small manufacturers and processors may exempt numerous manufacturers and processors of nanomaterials from such rules. Some stakeholders told us this exemption may be particularly limiting in the case of nanomaterials because much nanomaterial development is being done by small startup companies.

Moreover, the reporting rule that EPA intends to propose will not require periodic updates of the material reported.

EPA also collects data on chemicals through its Inventory Update Rule. Under this rule, EPA requires companies to regularly report certain information, including production volume and use information for chemicals they produce in quantities over 25,000 pounds.[32] This reporting threshold is intended to capture information on chemicals that account for most of the total U.S. production volume covered by TSCA. EPA has not adjusted this threshold to capture the production of nanomaterials, and thus EPA may be missing the opportunity to collect important information on nanomaterials under this rule.

Under TSCA, EPA can also issue rules that require chemical companies to test chemicals for their health and environmental effects. To require testing, EPA must find that a chemical (1) may present an unreasonable risk of injury to human health or the environment or (2) currently is or will be produced in substantial quantities and that either (a) there is or may be significant or substantial human exposure to the chemical or (b) the chemical enters or may reasonably be anticipated to enter the environment in substantial quantities. EPA must also determine that there are insufficient data to reasonably determine or predict the effects of the chemical on health or the environment and that testing is necessary to develop such data. EPA officials told us they intend to propose a rule in December 2010 that would require companies to generate test data on the health effects of 15 to 20 different nanomaterials, including carbon nanotubes, nanoclays, and nano aluminum, and also on nanomaterials used in aerosol-applied products.[33] This information will help EPA correlate the properties of these materials with specific health effects, manage or minimize risk and exposure, and help EPA determine the need for additional testing of these nanomaterials, according to EPA. EPA officials told us they will be working with the National Institute for Occupational Safety and Health, the Occupational Safety and Health Administration, and the Consumer Product Safety Commission on this effort. However, as we have noted in a prior report, EPA has had difficulty in promulgating test rules in the past because, as described above, it must demonstrate that chemicals may pose certain health or environmental risks or meet volume and exposure thresholds before it can require companies to establish such risks through testing.[34] Because relatively little is currently known about the potential risks of nanomaterials and many of them have low production volumes, EPA may have similar difficulties in making the types of determinations necessary to promulgate a test rule for nanomaterials.

EPA Has Not Developed a Clear Process under FIFRA for Regulating Pesticides Containing Nanomaterials

FIFRA requires companies to obtain a registration in order to distribute or sell a pesticide. According to EPA, this authority extends to pesticides containing nanomaterials. EPA must register a pesticide if it determines, among other things, the pesticide will perform its intended function without unreasonable adverse effects on the environment.[35] Under FIFRA, EPA is authorized to require companies to submit or generate data that EPA needs to assess the risks of the pesticide. EPA may publish and periodically revise both data requirements and guidelines identifying the types of information it generally requires to assess pesticides for registration and the methods by which such data may be generated. According to EPA, the agency may, on a case-by-case basis, modify data requirements and guidelines for specific pesticides. In making its registration decision, EPA can allow the pesticide to be distributed and sold; allow it to be distributed and sold under certain conditions, such as the need to develop further information; or prohibit its distribution and sale altogether. However, according to the agency, EPA's current guidelines do not require companies to specify whether their pesticides contain nanoscale materials.

Officials told us that since 2007 they have received a few applications for registration of various nanosilver pesticide preparations. EPA officials told us that some of the companies that have submitted registration applications for nanopesticides have told EPA that the pesticide includes nanomaterials, while in other cases EPA told us they were able to determine the pesticide contained nanomaterials from the manufacturing processes. However, EPA officials told us they registered at least one pesticide since 2007 without being aware that it contained nanomaterials. A group of environmental and consumer organizations has identified 260 products currently on the market that claim to contain nanosilver. This group contends these products should be regulated as pesticides due to the antimicrobial effects of nanosilver, but that these products are not registered with EPA under FIFRA.[36] Because applicants do not have to identify whether their pesticidal product contains nanomaterials, EPA may not know that certain pesticides contain nanomaterials, and these pesticides may be entering the market without EPA specifically considering the potential risks their nanomaterials may pose.

EPA officials told us that if a company replaces a conventionally sized active ingredient in a pesticide with a nanoscale version of that ingredient, it is mandatory for the company to amend its registration. Officials also noted, however, that the agency's position on this point needs to be made explicit to the regulated community and such a clarification could be made in EPA guidance. According to stakeholders, manufacturers of nanopesticides are required to obtain an amended registration in such a circumstance even without new EPA guidance explicitly requiring it since the registration requirement is based not only on questions of chemical identity, but also on claims made about the pesticide; its composition; and its chemistry, toxicology, and other information. However, until EPA makes the requirement to obtain an amended registration for pesticides that substitute a nanoscale ingredient for a conventionally sized ingredient clear, such pesticides may be re-engineered to include nanomaterials without EPA's knowledge and review.

EPA Believes It Has the Authority to Regulate Nanomaterials under Air, Water, and Waste Statutes but Technology-related Limitations and Volume-based Regulatory Thresholds Present Regulatory Challenges

According to EPA officials and stakeholders, the agency can regulate nanomaterials as it regulates other pollutants and waste under the Clean Air Act, Clean Water Act, and RCRA, as well as undertake cleanups of nanomaterials under CERCLA. Nanomaterials do not pose the same definitional difficulties under the air, water, and waste statutes as they do under TSCA and FIFRA because pollutants and wastes are defined by their effects on humans and the environment rather than by their composition. For example, EPA can list a nanomaterial as a hazardous air pollutant if the agency can establish that the nanomaterial may present a threat of adverse human health effects.[37] Similarly, EPA can list a nanomaterial as a toxic water pollutant if exposure to the nanomaterial causes death, disease, and genetic mutations, among other effects. Similarly under RCRA, a material is characterized as a hazardous waste if it is specifically listed as hazardous waste by EPA or it demonstrates any of four hazardous characteristics—ignitability, corrosivity, reactivity, or toxicity—based on testing or the knowledge of the manufacturer or processor that generated the waste. Finally, under CERCLA, a material is characterized as a hazardous substance if it is deemed hazardous under CERCLA, RCRA, the Clean Water Act, the Clean Air Act, or TSCA. EPA can designate additional substances as hazardous under CERCLA if their release may present substantial danger to the public health or welfare or the environment.

According to EPA officials and stakeholders, the agency faces technical challenges to enforcing certain statutory provisions for nanomaterials in air, water, and waste. For example, some stakeholders told us that because fine particulates (particulates under 2.5 micrometers in diameter) are already defined as a conventional air pollutant under the Clean Air Act,[38] EPA could apply this conventional air pollutant standard to nanomaterials. However, EPA officials told us that while they could regulate nanomaterials under this standard, they do not yet have the technology needed to monitor particles of this size to enforce the standard. According to EPA and stakeholders, the agency may need to reassess how it measures pollutants under the Clean Air Act with respect to nanomaterials. This is because given the relatively small weight associated with nanomaterials, EPA may need to count particles or measure their surface area rather than weigh them, as the current air pollutant standard calls for.

Similarly, according to some stakeholders, in order to enforce any technology-based effluent limitations for nanomaterials established under the Clean Water Act in the future, EPA would need to identify technology that can reliably and economically measure these materials in effluents, which it does not currently have.[39] Similarly, EPA may face challenges in regulating nanomaterials in waste under RCRA because the tests used to establish the hazards of waste in general may be inadequate to characterize the hazards of nanomaterials. For example, according to some stakeholders, to the extent that nanoparticles behave in significantly different ways than larger-scale particles in soil, groundwater, and drinking water, EPA's assumptions under current testing procedures may not fully assess how toxic wastes containing nanomaterials might affect groundwater.

In regulating nanomaterials, EPA also faces challenges attributable to volume-based thresholds and special conditions, such as waste coming from households, that trigger application of air, water, and waste laws and regulations. For example, EPA exempts household waste from RCRA hazardous waste regulation because it is impractical to regulate individual households. Moreover, EPA officials told us that landfill liners, as described in EPA's criteria for municipal solid waste landfills under RCRA, are sufficient to handle the small amounts of hazardous waste that end up in municipal landfills as a result of the household hazardous waste exemption. However, some stakeholders argue that until the risks of nanomaterials are better understood, it will not be known whether the landfill liners are sufficient to address the potential risks of nanomaterials that might be present in household waste. An example of a volume-based threshold issue arises under the Emergency Planning and Community Right to Know Act.[40] EPA has set thresholds in the regulations implementing hazardous chemical inventory reporting requirements under these provisions that may not establish a threshold that is appropriate for nanomaterials. For example, the regulations include a default inventory reporting threshold for releases of 500 pounds for extremely hazardous substances and releases of 10,000 pounds for other hazardous chemicals. Stakeholders question whether these thresholds may be too high in the context of nanomaterials. EPA can set the thresholds lower than the defaults and has, for example, reduced the default threshold for some specific extremely hazardous substances to 1 pound. However, it has not yet done so for any nanomaterials.

In addition to the challenges that EPA faces in regulating nanomaterials under air, water, and waste statutes, the agency may also be missing certain opportunities for gathering information on nanomaterials under the Clean Water Act. For example, EPA may not be collecting all available data on nanomaterials discharged into water. EPA has authority under the Clean Water Act to require owners or operators of facilities discharging pollutants to keep records, report information, monitor and sample discharges, and provide other information that EPA may reasonably require to carry out the act. The act also gives EPA the authority to inspect facilities and review records.[41] According to stakeholders, at least one court has interpreted this authority broadly, upholding as reasonable an EPA permit requirement directing an applicant to disclose all toxic pollutants used or produced in the facility.[42] Thus, stakeholders pointed out that EPA was able to obtain information not only on toxic pollutants that were in fact being discharged from a facility, but on those that had the potential to be discharged as well. Stakeholders concluded that even if EPA cannot currently measure nanomaterial discharges or cannot impose monitoring requirements on facilities, the agency has the ability to obtain information on the potential for nanomaterial discharge by a facility.

Other National Authorities Are Collecting Information on Nanomaterials and Are Evaluating Their Legislation to Ascertain if Changes Are Needed

Australia and the United Kingdom have undertaken a voluntary approach to collecting information on nanomaterials while Canada plans to require companies to submit certain data. In contrast, the European Union collects data on all chemicals being produced at a certain volume as required by its basic chemicals legislation, which also includes nanomaterials. All of these entities are reviewing their existing legislation to determine the need for additional regulatory authority to specifically address nanomaterials.[43]

Australia Has Asked Companies to Voluntarily Provide Information on Nanomaterials and Is Currently Reviewing Comments on Proposed Legislative and Regulatory Changes

Australia's National Industrial Chemicals Notification and Assessment Scheme (NICNAS)—the government's regulatory body for chemicals—has issued two requests for companies to voluntarily provide information on nanomaterials but, like the U.S. experience, these requests have produced limited results. In February 2006, NICNAS issued a voluntary request for information from industry on the uses and quantities of nanomaterials being manufactured or imported into the country. Nanomaterials used exclusively in certain products, such as sunscreens and food additives, among others, do not fall within the scope of NICNAS and were consequently outside the request for information. Data requested included chemical and trade name, molecular formula, and estimates of total quantity imported or manufactured, and NICNAS did not request data on nanomaterial toxicity. Companies supplied information on 21 types of nanomaterials, 17 of which were available for commercial use. The largest group of nanomaterials reported was metal oxides, which are used in surface coatings, water treatment, cosmetics, and catalysts. In October 2008, Australia expanded the information requested in 2006 when it initiated a second request for information that targeted all manufacturers or importers of nanomaterials or products containing nanomaterials for commercial or research and development purposes. The second request was for companies to identify what data they have on their nanomaterials' toxicological properties, while not requiring the data be provided to NICNAS. The request also stipulated that no new data needed to be generated. Although information was due to NICNAS by the end of January 2009, the results of this request have not yet been made public.

In addition to collecting information, NICNAS announced in fall 2009 that it is reviewing Australia's legislative framework and administrative practices to ensure that any potential risks from nanomaterials are adequately identified and appropriately managed. A 2008 review by an Australian university determined that Australia's regulatory frameworks should be reviewed to ensure that the risks posed by nanotechnology are better managed.[44] The following are areas, among others, that were identified for review by the report.

- *Classification of nanomaterials as new or existing*. Uncertainty exists as to whether the nano-form of a chemical is considered new or an existing chemical under current legislation. The NICNAS new chemicals program—for chemicals not listed on the national inventory—currently applies to nanomaterials and allows for them to be assessed before commercial use. However, nanoscale versions of existing chemicals—chemicals already on the national inventory—can legally be introduced and used without notification to NICNAS.

- *Weight or volume.* Some Australian regulatory requirements are currently triggered by weight or volume. For nanomaterials, weight or volume thresholds may not be meaningful because current production levels of nanomaterials are too low to trigger the threshold and not enough is known about the appropriate threshold levels.
- *Risk assessment protocols.* It is uncertain whether risk assessment methods currently being employed by various regulatory agencies are suitable for goods that contain nanomaterials. Such uncertainties reduce confidence in the results of assessments.

To address these areas of concern, Australia's NICNAS has proposed a range of reforms, including removing nanomaterials from certain exemptions and potentially requiring nanomaterials based on conventionally scaled existing chemicals to go through the new chemicals review program. Public comment on these proposals closed on February 12, 2010, but the results of these comments have not yet been made public.

The United Kingdom Has Asked Companies to Voluntarily Report Certain Data on Nanomaterials and Is Currently Reviewing Whether Legislative Changes Are Needed

The United Kingdom launched a voluntary reporting scheme for nanomaterials in 2006 that targeted manufacturers, importers, and users and that also resulted in the collection of limited information. This effort focused on free nanomaterials—nanomaterials not enclosed in other materials—because they were identified as having greater potential for environmental exposure. Information requested included chemical identity; dimensions and shape; size range; predictions of surface area; uses; available toxicological data; and certain physical and chemical characteristics, such as water solubility, stability, and flammability. As of July 2007, the United Kingdom had only received nine responses to its voluntary reporting scheme.

Regarding legislation, the United Kingdom commissioned reviews of the adequacy of existing legislation for each of its key regulatory departments to assess whether current regulatory frameworks are adequate to address the potential risks posed by nanomaterials. In general, these reviews concluded that the current regulatory framework, while broadly sufficient, has the potential for nanomaterials to fall outside of regulatory controls in certain circumstances, such as regulations with production volume or mass thresholds developed in the context of macroscale materials. The review also found that certain consumer products containing nanomaterials may be found safe for consumer use, but that risk assessments may not consider the full product life cycle, including its disposal. Consequently, in June 2009, the United Kingdom recognized that there may be a need to adjust existing systems to create a more integrated approach to address risks from nanomaterials. The United Kingdom is currently considering these issues as it develops its strategy on nanotechnologies.

Canada Is Drafting a Requirement That Companies Provide Information on Nanomaterials and Plans to Review the Data Collected before Proposing Any Regulatory Changes

Canadian officials have proposed but have not implemented a one-time requirement for companies to provide information on nanomaterials produced in or imported into Canada. Canadian importers and manufacturers would be required to report their use of nanomaterials produced or imported in excess of 1 kilogram. In 2009, Canadian officials reported to the OECD that information required would include chemical and trade name; molecular formula; and any available information on the shape, size range, structure, quantity imported or manufactured, and known or predicted uses. Also required would be any available information on the nanomaterial's physical and chemical properties—such as solubility in water and toxicological data, among others. Under the proposal, companies could claim information as confidential, but regulators would publish a summary of information provided. Canada plans to use this information to help develop a regulatory framework for nanomaterials and to determine which information requirements would be useful for subsequent risk assessments. Canadian officials stated they originally hoped to issue this requirement in the spring of 2009 but could not predict when it would be implemented.

With regard to current law, a report prepared for the government of Canada in 2008 stated that Canada has no specific requirements for nanomaterials and is considering whether they are needed. However, Health Canada and Environment Canada—two agencies responsible for health and the environment—have taken the first steps in recognizing the potentially unique aspects of nanomaterials. These regulatory agencies are currently relying on existing authority delegated to them through legislation, such as the Canadian Environmental Protection Act, to address nanomaterials. Specifically, in June 2007, Environment Canada released a new substances program advisory announcing that nanomaterials will be regulated under the act's new substances notification regulations. Per this advisory, any nanomaterial not listed on Canada's chemical inventory—the Domestic Substances List—or with "unique structures or molecular arrangements" compared to their non-nano counterparts, requires a risk assessment. A review panel of the Canadian Academies found that, while it is not necessary to create new regulatory mechanisms to address the unique challenges presented by nanomaterials, the existing regulatory mechanisms could and should be strengthened in a variety of ways, such as by creating a specific classification for nanomaterials and by reviewing the regulatory triggers that prompt review of the health and environmental effects.

The European Union Is Considering Revising Its Chemicals Legislation to Better Address Nanomaterials, and Is Requiring Labeling of Nanomaterials in Certain Products

The European Union passed its chemical legislation in 2007, known as Regulation, Evaluation and Authorization of Chemicals (REACH),[45] under which the European Union generally collects information on all chemicals. However because REACH requirements apply to chemicals with a production volume of greater than 1 metric ton per year, some stakeholders have expressed concern that the provisions of REACH will not identify the risks of most nanomaterials because companies do not produce these materials at this level or volume. Because of this concern, the European Union is reviewing whether the provisions of REACH need to be modified to take into consideration the unique properties of nanomaterials by, for example, adjusting the volume-based thresholds. This review is ongoing, according to official EU reports, and is not scheduled for completion until 2012.

In addition to efforts under REACH, the European Union has developed a regulation to require labeling on certain types of products containing nanomaterials. For example, a European Union Cosmetics Regulation will require cosmetic products that contain nanoscale ingredients to be labeled as such. The regulation would also require the manufacturers of new cosmetic products containing nanomaterials to notify regulators and provide them with certain safety information. Manufacturers of products containing nanoscale ingredients already being sold in the European Union also would have to notify regulators and submit certain safety information. In addition, the regulation requires all nanomaterial ingredients be clearly indicated in the list of ingredients and the names of such ingredients shall be followed by the word "nano" in brackets. The regulation also calls for the European Commission to compile a publicly available catalogue of all nanomaterials used in cosmetic products placed on the market, including those used as colorants, UV filters, and preservatives. Although this regulation was published in November 2009, its provisions are not scheduled to go into effect until July 2013.

In addition to the Cosmetics Regulation, the European Union has also begun to regulate nanomaterials in food. Specifically, in January 2010, revised regulations on food additives went into effect. The regulations clarify that when there is a change in the particle size of a previously approved food additive, a new approval is required before the additive goes to market. The European Union is also considering an update to its regulations on novel foods—foods or ingredients not widely consumed by people prior to 1997—that includes measures to regulate manufactured nanomaterials in food. Specifically, the proposed update would require that all foods containing manufactured nanomaterials undergo premarket authorization.

Some State and Local Governments Have Begun to Address the Risks of Nanomaterials

Some U.S. states and localities have begun to address the potential risks from nanomaterials by, for example, issuing requests for information. Specifically, in January 2009, California required companies that manufacture or import carbon nanotubes into the state submit certain readily available data on these materials to the California Department of Toxic Substances Control by January 22, 2010. California officials told us that carbon nanotubes are an important category of emerging nanomaterials for which data on toxicity, physiochemical properties, and environmental fate and transport are largely unavailable. California posted the 22 responses it received on its Web site, as well as the names of companies that failed to respond. In addition, California environmental officials said they are now considering whether to conduct additional information requests on nanoscale forms of metal oxides, including nano aluminum oxide, nano silicon dioxide, nano titanium dioxide, and zinc oxide, as well as nanosilver, nano zerovalent iron, and nano cerium oxide. According to stakeholders we spoke with, environmental officials in other states have also considered similar information requests. For example, in 2009, some Wisconsin state legislators called for a study on the feasibility of creating a nanotechnology registry and the development of subsequent legislation.

In addition to states, some municipalities have considered collecting information on nanomaterials. For example, in December 2006, the City of Berkeley, California, issued a hazardous materials ordinance that requires companies to report the manufacture or use of nanomaterials. According to stakeholders we spoke with, this was the first time a U.S. city took such an approach. Berkeley's ordinance requires that facilities that manufacture or use nanoparticles submit a separate written disclosure of the material's known toxicology and how the facility will safely handle, monitor, contain, dispose, track, and mitigate the risks of such materials. Cambridge, Massachusetts, also considered implementing a similar ordinance but has not done so yet.

Several state environmental officials told us they have considered whether their states' current regulations provide enough authority to address the risks of nanomaterials. For example, environmental officials in California told us they planned to review the data gathered under their requests for information to determine if additional action is needed. According to a report issued by the Environmental Council of the States, [46] other states are taking some preliminary actions with regard to nanomaterials. Specifically,

- Maine officials developed an Air Toxics Priority List in July 2007 that includes particulate matter from nanotechnology,
- the Massachusetts Department of Environmental Protection identified nanomaterials as an emerging contaminant of concern and established an Interagency Nanotechnology Committee,
- the Washington State Department of Ecology considers nanomaterials to be an emerging contaminant of concern and has revised its manual for hazardous waste inspectors to include specific information on nanomaterials, and

- Pennsylvania and South Carolina have identified nanoparticles as contaminants of concern.

The report identified nanomaterials, among other substances, as emerging contaminants of concern.[47] The report specifically requested that federal agencies consider nanomaterials as a special class of emerging contaminants due to properties that may make them behave in ways that conventional-scale contaminants do not. In addition, the report identified a number of states that are taking some preliminary actions with regard to nanomaterials.

CONCLUSIONS

The use of nanomaterials in products is growing faster than our understanding of the risks these materials pose to human health and the environment. While EPA has taken steps to improve our understanding of these risks, such as by asking companies to voluntarily provide information on the nanomaterials they produce, the information gathered through these efforts has been limited and does not provide a strong foundation for understanding the increasing potential for exposure to these materials as their uses become more prevalent. EPA has taken some regulatory action with regard to nanomaterials under TSCA and has developed plans to take further action with regard to information collection and testing of nanomaterials. However, these changes have not yet gone into effect and products may be entering the market without EPA review of available information on their potential risk. Moreover, although EPA requires chemical companies to periodically provide certain information on many of the chemicals currently in commerce, EPA has not extended this requirement to nanomaterials. Thus, EPA may be missing the opportunity to gather some additional information on nanomaterials from the regulated community. Furthermore, although EPA is taking steps to regulate pesticides containing nanomaterials, it has not clearly stated this to manufacturers, and the current data requirements do not require companies to specify whether any materials in their pesticides are nanoscaled.

EPA also may be missing the opportunity to gather some additional information on potential discharges of nanomaterials from the regulated community. We acknowledge that EPA faces technical challenges in its research and regulatory efforts caused in part by a lack of tools and models to help generate information on the potential risks; however, better use of existing environmental statutes, such as the Clean Water Act, may enable EPA to collect useful information on nanomaterials.

RECOMMENDATIONS FOR EXECUTIVE ACTION

We recommend that the Administrator of EPA, take the following three actions:

- Complete its plan to issue a Significant New Use rule for nanomaterials.
- Modify FIFRA pesticide registration guidelines to require applicants to identify nanomaterial ingredients in pesticides.
- Complete its plan to clarify that nanoscale ingredients in already registered pesticides, as well as in those products for which registration is being sought, are to be reported to EPA and that EPA will consider nanoscale ingredients to be new.
- In addition, the Administrator of EPA should make greater use of the agency's authorities to gather information under existing environmental statutes. Specifically, EPA should
- complete its plan to use data gathering and testing authorities under TSCA to gather information on nanomaterials, including production volumes, methods of manufacture and processing, exposure and release, as well as available health and safety studies; and
- use information-gathering provisions of the Clean Water Act to collect information about potential discharges containing nanomaterials.

Finally, the Administrator of EPA should consider revising the Inventory Update Rule under TSCA so that it will capture information on the production and use of nanomaterials and so that the agency will receive periodic updates on this material.

APPENDIX I: OBJECTIVES, SCOPE, AND METHODOLOGY

Our objectives for this review were to (1) identify examples of current and potential uses of nanomaterials, (2) determine what is known about the potential human health and environmental risks from nanomaterials, (3) specifically assess actions the Environmental Protection Agency (EPA) has taken to better understand and regulate nanomaterials as well as its authorities to do so, and (4) identify approaches that selected national authorities have taken to address the risks associated with nanomaterials. In addition, you asked us to identify any U.S. states and localities that may have begun to address risks from nanomaterials.

To identify examples of current and potential uses of manufactured nanomaterials, we analyzed documents and reports created by stakeholders, including synthesis studies, databases of nanotechnology-related products, and Web sites that compiled and analyzed nanotechnology-related products from various sources. We identified the documents and reports (1) through interviews with knowledgeable stakeholders, (2) through open source research, and (3) from a literature search. Because of the dynamic nature of nanotechnology, we used only documents published since 2005. We also sought reports that sorted the current and potential uses of nanomaterials into broad categories, so that our report would not exclude any major industry sectors. We analyzed the information, compared the sets of industry sectors used in various reports to each other, and created a list of eight industry sectors that in our estimation reflected the breadth and depth of the commercial market for products enabled by nanomaterials. We selected specific examples within each sector for further analysis. Because assembling a comprehensive catalog of uses would be difficult in an evolving, dynamic industry, our list of examples is not comprehensive but rather was selected in a manner that allowed us to convey the wide spectrum of materials in current use, or which could be in use in the future, across a large range of products. In addition, we interviewed cognizant agency officials from the top six agencies conducting nanotechnology-related research. These six agencies accounted for over 95 percent of federal nanotechnology research reported in fiscal year 2009.[48] We also interviewed knowledgeable stakeholders, including officials from the National Nanotechnology Initiative, the Woodrow Wilson International Center for Scholars' Project on Emerging Nanotechnologies, Lux Research—an independent research firm that conducts market analysis of nanotechnology, among other things—and the NanoBusiness Alliance—a nanotechnology related business association. To identify knowledgeable stakeholders, we used an iterative process, often referred to as "snowball sampling," in which we asked our initial interviewees to identify others we should talk to, and we selected for interviews those who would provide us with a broad range of perspectives on the current and potential uses of nanomaterials.

To determine what is known about the potential human health and environmental risks of nanomaterials, we reviewed documents that had been published by peer-reviewed journals, government agencies, and international nonprofit organizations. In conducting this review, we searched databases, asked knowledgeable stakeholders to identify relevant studies, and reviewed studies from article bibliographies to identify additional sources of information on the potential risks. Because of the importance of using the most current risk-related research, the team used only documents published since 2005. Of the over 700 documents we identified published between 2005 and 2010, we narrowed our review to 140. Of these, we selected 20 for more detailed analysis. We selected these documents in large part because they provided a synthesis of available research related to nanomaterials risks and they covered a variety of nanomaterials. To assess the credibility, reliability, and methodological soundness of these publications, a senior GAO technology analyst reviewed each of the publications and considered such factors as the bibliographies of evidence cited and the location of where the articles were published. We did not examine the references cited by these studies as part of our analysis. We concluded that all 20 reviews were sufficiently reliable for the purposes of this report. For the purposes of this review, all the documents, studies, and syntheses we reviewed will be referred to in our report as "studies." We also spoke with a variety of knowledgeable stakeholders representing government, industry, academia, nongovernmental organizations, and the regulatory community. These knowledgeable stakeholders were also selected using a snowball sampling method.

To assess actions EPA has taken to better understand and regulate nanomaterials and its authorities to do so, we analyzed selected laws and regulations, including the Toxic Substances Control Act of 1976; the Federal Insecticide, Fungicide, and Rodenticide Act; the Clean Air Act; the Clean Water Act; the Resource Conservation and Recovery Act; and the Comprehensive Environmental Response, Compensation, and Liability Act. We also reviewed data and reports on EPA's Nanoscale Materials Stewardship Program, which EPA developed to encourage companies to voluntarily develop and submit information to EPA on the characteristics of nanomaterials. We interviewed and obtained documentation from agency officials responsible for implementing these laws in EPA's Office of Air and Radiation, Office of Pollution Prevention and Toxic Substances, Office of Pesticide Programs, Office of Solid Waste and Emergency Response, and Office of Water. We also interviewed and obtained documentation from staff in EPA's Office of Research and Development. Furthermore, we consulted with knowledgeable stakeholders and legal experts to obtain their perspectives on EPA's available authorities to regulate nanomaterials.

APPENDIX II: COMMENTS FROM THE ENVIRONMENTAL PROTECTION AGENCY

UNITED STATES ENVIRONMENTAL PROTECTION AGENCY
WASHINGTON, D.C. 20460

MAY 04 2010

OFFICE OF CHEMICAL SAFETY AND POLLUTION PREVENTION

Anu Mittal, Director
Natural Resources and Environment
U.S. Government Accountability Office
441 G Street, NW, Room HQ 2T31
Washington, D.C. 20548

Dear Ms. Mittal:

Thank you for the opportunity to review the U.S. Government Accountability Office's (GAO's) draft report entitled *Nanotechnology: Nanomaterials Are Widely Used in Commerce, but EPA Faces Challenges in Regulating Risk* (GAO-10-549). We appreciate and concur with GAO's recommendations, which are consistent with the Agency's approach to effectively managing nanoscale materials.

We note GAO's acknowledgement that there are unanswered questions about the potential risks of nanoscale materials to human health and the environment. The same special properties that make nanoscale materials useful are also properties that may cause some nanoscale materials to pose potential risks to humans and the environment. At this point, not enough information exists to fully assess these risks. EPA will need a sound scientific basis for assessing and managing any unforeseen future impacts resulting from the introduction of nanoscale materials into the environment, as well as informing material design and use decisions that avoid or reduce risk. A challenge for environmental protection is to help fully realize the societal benefits of nanotechnology while identifying and minimizing any adverse impacts to humans or ecosystems from exposure to nanoscale materials. The growing diversity and complexity of the types and uses of nanoscale materials available and being developed presents challenges in evaluating risks associated with the manufacture and use of these materials.

As noted in the GAO report, this understanding will come from environmental research and development activities. EPA is heavily involved in efforts to understand the potential risks to humans, wildlife, and ecosystems from exposure to nanomaterials. Through innovation and discovery, scientists are studying the unique properties of nanomaterials, determining their potential impacts, and developing approaches to evaluate and prevent any risks. They are also exploring how nanomaterials can be used effectively to clean up contaminants released into the environment. With the use of nanotechnology in the consumer and industrial sectors expected to increase significantly in the future, nanotechnology offers society the promise of major benefits. The challenge for environmental protection is to ensure that, as nanomaterials are developed and used, unintended consequences of exposures to humans and ecosystems are prevented or minimized.

The recommendations contained in the GAO draft report are consistent with the Agency's approach to effectively managing nanoscale materials and we accept them. Below are comments on each of the recommendations found in the draft report.

EPA Response to GAO's Recommendations for Executive Action

We recommend the Administrator, EPA, take the following actions:

- Complete its plan to issue a Significant New Use Rule for nanomaterials.

  EPA agrees. EPA will continue to issue SNURs for nanoscale materials that are new chemical substances on a case-by-case basis, as appropriate, and intends to propose a SNUR for nanoscale materials that are existing chemical substances by December 2010.

- Modify FIFRA pesticide registration guidelines to require applicants to identify nanomaterial ingredients in pesticides.

  EPA agrees and intends to clarify that, as part of the application for registration, applicants for pesticide registrations which contain nanomaterial ingredients need to specifically identify those ingredients.

- Complete its plan to clarify the FIFRA guidelines to make clear that already registered pesticides that have been reengineered to include nanomaterials need to obtain an amended registration.

  EPA agrees and is working on clarification of registrant's responsibilities under FIFRA with respect to nanomaterials.

In addition, GAO recommends the EPA Administrator should make greater use of its authorities to gather information under existing environmental statutes. Specifically, GAO recommends EPA should:

- Complete its plan to use data gathering authorities under the Toxic Substances Control Act (TSCA) to gather information on nanomaterials, including production volumes, methods of manufacture and processing, exposure and release, as well as available health and safety studies.

  EPA agrees and intends to propose a section 8(a) information gathering rule as described in the recommendation and also intends to propose a section 4 test rule.

- Use information gathering provisions of the Clean Water Act to collect information about potential discharges containing nanomaterials.

  The Agency agrees that collecting information about discharges is a critical component of understanding potential environmental risks. EPA's Office of Research and Development, and others, is conducting research to determine whether nanomaterials may enter the water in forms and levels of concern, as well as how to detect and monitor nanomaterials in effluents and aquatic systems. Once we have these capabilities, EPA will consider whether new reporting requirements should be applied to companies who may be discharging nanomaterials into the environment, including under the Clean Water Act.

- Finally, the EPA Administrator should consider revising the Inventory Update Rule under TSCA so that it will capture information on the production and use of nanomaterials and so that the Agency will receive periodic updates on this material.

EPA agrees and will consider proposing periodic reporting under the Inventory Update Rule for nanoscale materials.

Again, we appreciate the opportunity to review and comment on this draft report. Should you have any questions or concerns regarding this response, please contact Bob Trent, EPA's GAO Liaison Team Lead, at 202-566-0983.

Sincerely,

Stephen A. Owens
Assistant Administrator

To determine which national authorities had recently addressed nanomaterials, we interviewed knowledgeable stakeholders, including EPA officials who participated in working groups within the Organisation for Economic Co-operation and Development to identify candidate national authorities. We selected a judgmental sample of four countries for our review based on the following criteria: (1) EPA officials agreed that these countries have robust environmental regulations that were comparable to US regulations and (2) the countries had recently taken action with regard to nanomaterials, including considering to regulate nanomaterials. Based on this, we selected Australia, Canada, the United Kingdom, and the European Union. To identify the approaches these national authorities have used to address the potential risks associated with nanomaterials, we analyzed these authorities' laws and regulations that would be applicable to regulating nanomaterials, reviewed reports that other organizations had conduced of these countries' laws as they pertain to nanotechnology, and supplemented our understanding with interviews with knowledgeable stakeholders and legal experts.

To identify any states or local governments that may be taking action with regard to nanomaterials, we interviewed with knowledgeable stakeholders including EPA officials, representatives from environmental organizations, and the Environmental Council of States—a nonpartisan association of state environmental officials. We collected and analyzed documentation on these activities and supplemented our analysis with interviews with selected state officials.

Related GAO Reports

Food Safety: FDA Should Strengthen Its Oversight of Food Ingredients Determined to Be Generally Recognized as Safe (GRAS). GAO-10-246. Washington, D.C.: February 3, 2010.

Chemical Regulation: Observations on Improving the Toxic Substances Control Act. GAO-10-292T. Washington, D.C.: December 2, 2009.T

High-Risk Series: An Update. GAO-09-271. Washington, D.C.: January 22, 2009.

Federal Research: Opportunities Exist to Improve the Management and Oversight of Federally Funded Research and Development Centers. GAO-09-15. Washington D.C.: October 8, 2008.

Highway Safety: Foresight Issues Challenge DOT's Efforts to Assess and Respond to New Technology-Based Trends. GAO-09-56. Washington, D.C.: October 3, 2008.

Nanotechnology: Accuracy of Data on Federally Funded Environmental, Health, and Safety Research Could Be Improved. GAO-08-709T. Washington, D.C.: April 24, 2008

Nanotechnology: Better Guidance Is Needed to Ensure Accurate Reporting of Federal Research Focused on Environmental, Health, and Safety Risks. GAO-08-402. Washington, D.C.: March 31, 2008.

Chemical Regulation: Comparison of U.S. and Recently Enacted European Union Approaches to Protect against the Risks of Toxic Chemicals. GAO-07-825. Washington, D.C.: August 17, 2007.

21st Century Challenges: Reexamining the Base of the Federal Government. GAO-05-325SP. Washington, D.C.: February 2005.

End Notes

[1] EPA is one of four key agencies that administer laws that regulate manufactured nanomaterials depending on how they are used. The other regulatory agencies include the Consumer Product Safety Commission, the Department of Health and Human Services' Food and Drug Administration, and the Department of Labor's Occupational Safety and Health Administration. We did not review these other agencies' regulatory authorities as part of this work.

[2] In addition, EPA has authority under the Federal Food, Drug, and Cosmetic Act to establish tolerances or exemptions for the requirement of a tolerance for pesticide residues that remain in food. Food is considered adulterated if, amongst other conditions, it contains any residue of a pesticide chemical for which there is no tolerance or exemption or which exceeds any established tolerance.

[3] These agencies are the Department of Defense, the Department of Energy, EPA, the Department of Health and Human Services' National Institutes of Health, the Department of Commerce's National Institute of Standards and Technology, and the National Science Foundation. These six agencies accounted for over 95 percent of federal nanotechnology research reported in fiscal year 2009.

[4] Green chemistry, also known as sustainable chemistry, is the design of chemical products and processes that reduce or eliminate the use or generation of hazardous substances. Green chemistry can be applied across the life cycle of a chemical product, including its design, manufacture, and use.

[5] Exercising foresight consists of basing policies on an understanding of forces shaping the future. In this context, a potentially significant trend is one that, although somewhat uncertain, may substantially affect progress toward basic goals across a time horizon more than 5 years forward.

[6] GAO, *Highway Safety: Foresight Issues Challenge DOT's Efforts to Assess and Respond to New Technology-Based Trends*, GAO-09-56 (Washington, D.C.: Oct. 3, 2008).

[7] GAO, *21st Century Challenges: Reexamining the Base of the Federal Government*, GAO-05-325SP (Washington, D.C.: February 2005).

[8] Rejeski, David, and Carly Wobig. 2002. Long-term Goals for Governments. Foresight 4, no. 6:14–22.

[9] The Wilson Center is a nonpartisan research institution established by an act of Congress in 1968 and supported by public and private funds. The products in the Wilson Center's consumer products database are products identified as containing nanomaterials by their manufacturers or another source, which can be readily purchased by consumers, and for which the nanomaterials-based claims for the product appear reasonable. The NanoBusiness Alliance, an industry association representing the nanotechnology business community, estimates that thousands of additional products using nanomaterials are not publicized by their manufacturers. These products would therefore most likely not be counted in databases like the Wilson Center's database.

[10] Carbon nanotubes are basically tubes that consist of rolled-up sheets of graphite. These materials have novel properties, including extraordinary strength and unique electrical conductivity.

[11] Cerium oxide additives are already in use on a large scale in bus fleets in a number of countries including the United Kingdom, but their sale is not currently authorized in the United States.

[12] A catalyst is a substance that increases the rate of a chemical reaction without being consumed by the reaction.

[13] A polymer is a material made of long, chain-like molecules.

[14] An ion is an atom or group of atoms that bears one or more positive or negative electrical charges.

[15] A quantum dot is a nano-sized crystal that efficiently absorbs light and emits either photons or electrons.

[16] A terabyte is about 1 trillion bytes or about 1,000 gigabytes.

[17] Photoelectrolysis is the splitting of water into hydrogen and oxygen using light energy.

[18] The Food and Drug Administration is generally responsible for overseeing the safety of color additives and foods, including food additives and dietary supplements, as well as for safety of food packaging.

[19] Photocatalysis is the acceleration of a photoreaction in the presence of a photocatalyst.

[20] The Food and Drug Administration is generally responsible for overseeing the safety and effectiveness of drugs and devices for humans and animals, and of biological products for humans.

[21] The Food and Drug Administration is generally responsible for overseeing the safety of cosmetics. In addition, the U.S. Consumer Product Safety Commission is responsible for protecting the public from unreasonable risks of serous injury or death from more than 15,000 types of consumer products, including some that may be manufactured with nanomaterials.

[22] The routes of exposure listed are generally for incidental or consumer exposures to nanomaterials. For medical applications, the primary route of exposure is intravenous.

[23] Some consumer products containing edible nanomaterials are available. Consumers may now purchase food containing nanomaterials such as prepared milkshakes containing nanoscale vitamins used to fortify the shakes.

[24] The Occupation Safety and Health Administration is responsible for ensuring the safety and health of workers by setting and enforcing standards and encouraging continual improvement in workplace safety and health.

[25] We selected six key statutes administered by EPA—TSCA, FIFRA, the Clean Air Act, the Clean Water Act, RCRA, and CERCLA—for the purpose of assessing actions EPA has taken to better understand and regulate the risks posed by nanomaterials as well as its authorities to do so. Also, as noted previously, EPA is one of four agencies that administers laws that regulate manufactured nanomaterials depending on how they are used. We did not review the other three agencies' regulatory authorities as part of this report, although we did identify nanomaterial uses that may be regulated by them.

[26] EPA has been conducting research in ultrafine particulate matter, particularly in the air. In this research, EPA defines ultrafine particles as those less than 100 nanometers, making them nanoscaled.

[27] The OECD is a forum for the governments of 30 developed countries to work together to address economic, social, and environmental issues.

[28] The 14 nanomaterials that the OECD has selected for further review are aluminum oxide, carbon black, cerium oxide, dendrimers, fullerenes, iron nanoparticles, multiwalled carbon nanotubes, nanoclays, polystyrene, silicon dioxide, silver nanoparticles, single-walled carbon nanotubes, titanium dioxide, and zinc oxide.

[29] Allotropes are different forms of the same element in which the atoms are arranged differently. For example, graphite and diamond are allotropes of carbon. Isotopes are different forms of the same element that have different atomic weights because they have different numbers of neutrons. For example, helium-3, which has two protons and one neutron in its nucleus, is an isotope of helium.

[30] The SNURs were originally issued as direct final rules—that is, they would go into effect without formal consideration of public comment after a certain period if EPA did not receive any adverse comments. Because EPA received a notice of intent to submit adverse comments, however, EPA withdrew the SNURs. When EPA proposed these rules again in November, it provided for a public comment period.

[31] EPA plans to propose this rule under section 8(a) of TSCA.

[32] Every 5 years, companies must report certain information on the production volume for chemicals they produced over 25,000 pounds at one location during that year. Companies

must also report additional use information on chemicals that they produce over 300,000 pounds at one location.

[33] EPA plans to propose this rule under section 4 of TSCA.

[34] GAO, *Chemical Regulation: Options Exist to Improve EPA's Ability to Assess Health Risks and Manage Its Chemical Review Program*, GAO-05-458 (Washington, D.C.: June 13, 2005).

[35] The phrase "unreasonable adverse effects on the environment" means (1) any unreasonable risk to man or the environment, taking into account the economic, social, and environmental costs and benefits of the use of any pesticide, or (2) a human dietary risk from residues that result from a use of a pesticide in or on any food inconsistent with the standard for tolerance under the Federal Food, Drug, and Cosmetic Act. 7 U.S.C. § 136(bb) (2006).

[36] In November 2008, a group of environmental and consumer organizations filed a petition asking EPA to regulate products containing nanosilver as pesticides. Petitioners included the International Center for Technology Assessment, the Center for Food Safety, Friends of the Earth, Greenpeace, the Center for the Study of Responsive Law, and the Consumers Union.

[37] EPA may promulgate a rule designating a given material as a hazardous air pollutant if the material presents, or may present, through inhalation or other routes of exposure, a threat of adverse human health effects (including carcinogenicity, mutagenicity, neurotoxicity, reproductive dysfunction, or acute or chronic toxicity) or adverse environmental effects whether through ambient concentrations, bioaccumulation, deposition, or otherwise. 42 U.S.C. § 7412(b)(2) (2006).

[38] A conventional air pollutant is one that causes or contributes to air pollution that may reasonably be anticipated to endanger public health or welfare. There are five other conventional air pollutants in addition to particulates: they are ground-level ozone, carbon monoxide, sulfur oxides, nitrogen oxides, and lead.

[39] An effluent limitation is a restriction on the discharge of pollutants from, for example, a factory, into the waters of the United States.

[40] Under this act, covered facilities must submit an emergency and hazardous chemical inventory form to (a) the appropriate local emergency planning committee; (b) the state emergency response commission; and (c) the fire department with jurisdiction over the facility. 42 U.S.C. § 11022(a) (2006).

[41] 33 U.S.C. § 1318 (2006).

[42] *NRDC v. EPA*, 822 F.2d 104, 119 (D.C. Cir. 1987).

[43] We selected a judgmental sample of four national authorities for our review, based on criteria such as countries that have recently taken action with regard to nanomaterials.

[44] Monash University, *Review of the Possible Impacts of Nanotechnology on Australia's Regulatory Frameworks* (May 2008).

[45] REACH's requirements are being phased in and will not be in full force until 2018.

[46] The Environmental Council of the States is the national non-profit, non-partisan association of state and territorial environmental agency leaders.

[47] Environmental Council of the States. *State Experiences with Emerging Contaminants: Recommendations for Federal Action,* January 2010.

[48] These agencies are the Department of Defense, the Department of Energy, EPA, the Department of Health and Human Services' National Institutes of Health, the Department of Commerce's National Institute of Standards and Technology, and the National Science Foundation.

In: Nanotechnology Considerations for the EPA... ISBN: 978-1-61761-817-8
Editors: Scott J. Castiglione

Chapter 2

NANOTECHNOLOGY AND ENVIRONMENTAL, HEALTH, AND SAFETY: ISSUES FOR CONSIDERATION

John F. Sargent
Science and Technology Policy

SUMMARY

Nanotechnology—a term encompassing nanoscale science, engineering, and technology—is focused on understanding, controlling, and exploiting the unique properties of matter that can emerge at scales of one to 100 nanometers. A key issue before Congress regarding nanotechnology is how best to protect human health, safety, and the environment as nanoscale materials and products are researched, developed, manufactured, used, and discarded. While the rapidly emerging field of nanotechnology is believed by many to offer significant economic and societal benefits, some research results have raised concerns about the potential adverse environmental, health, and safety (EHS) implications of nanoscale materials.

Some have described nanotechnology as a two-edged sword. On the one hand, some are concerned that nanoscale particles may enter and accumulate in vital organs, such as the lungs and brains, potentially causing harm or death to humans and animals, and that the diffusion of nanoscale particles in the environment might harm ecosystems. On the other hand, some believe that nanotechnology has the potential to deliver important EHS benefits such as reducing energy consumption, pollution, and greenhouse gas emissions; remediating environmental damage; curing, managing, or preventing diseases; and offering new safety-enhancing materials that are stronger, self-repairing, and able to adapt to provide protection.

Stakeholders generally agree that concerns about potential detrimental effects of nanoscale materials and devices—both real and perceived—must be addressed to protect and improve human health, safety, and the environment; enable accurate and efficient risk assessment, risk management, and cost-benefit trade-offs; foster innovation and public confidence; and ensure that society can enjoy the widespread economic and societal benefits that nanotechnology may offer. Congressionally-mandated reviews of the National Nanotechnology Initiative (NNI) by the National Research Council and the President's Council of Advisors on Science and Technology have concluded that additional research is required to make a rigorous risk assessment of nanoscale materials.

INTRODUCTION

Nanotechnology—a term encompassing nanoscale science, engineering, and technology—is focused on understanding, controlling, and exploiting the unique properties of matter that can emerge at scales of one to 100 nanometers.[1] These properties are believed by many to offer substantial economic and societal benefits.

A key issue before Congress regarding nanotechnology is how best to protect human health, safety, and the environment as nanoscale materials and products are researched, developed, manufactured, used, and discarded. While the rapidly emerging field of nanotechnology is believed by many to offer significant economic and societal benefits, some research results have raised concerns about the potential environmental, health, and safety (EHS) implications of nanoscale materials. Potential tools the Federal government might use to address these concerns include research and development, regulation, and international engagement.

Some of the properties of nanoscale materials (e.g., small size, high surface area-to-volume ratio) that have given rise to great hopes for beneficial applications have also given rise to concerns about their potential adverse implications for the environment, and human health and safety.[2] There are more than 600 nanotechnology products reportedly commercially available,[3] and with this number of products concerns have been raised about the health and safety of the scientists working with nanoscale materials, workers who manufacture the products, consumers who use the products, and members of the general public who may be exposed to nanoparticles, as well as the environmental impact of nanomanufacturing processes and the use and disposal of nanotechnology products.

Nanoscale particles can result from a variety of different processes. While nanoscale particles can occur naturally (e.g., some particles produced by forest fires, sea spray, volcanoes) and as an incidental by-product of human activities (e.g., some particles contained in welding fumes, diesel exhaust, industrial effluents, cooking smoke), EHS concerns have focused primarily on nanoscale materials that are intentionally designed and produced, often referred to as engineered nanomaterials.

Issues surrounding the potential EHS implications of nanotechnology emerged with the launch in 2000 of the National Nanotechnology Initiative (NNI). The NNI is a multi-agency federal effort to coordinate and expand federal nanotechnology research and development (R&D) efforts. Between FY2001 and FY2009, the federal government invested $9.9 billion in nanotechnology R&D, including approximately $1.5 billion in FY2009. Many governments around the world have followed the U.S. lead and established their own national nanotechnology programs. The private sector has invested heavily as well. Global nanotechnology R&D investments—public and private—are estimated to have totaled $12.4 billion in 2006 alone.[4]

Such large investments and intensified efforts to capitalize on these public and private investments have caused some observers (as detailed later in this report) to suggest that there is insufficient information about the potential effects nanotechnology products and manufacturing processes may have on human health, safety, and the environment. They assert a variety of uncertainties, including: how nanoscale particles might be transported in air, water, and soil; how they might react with the environment chemically, biologically, or through other processes; how they might be distributed and deposited; and whether they might accumulate in plants or animals.

Others express the view that concerns about nanotechnology EHS implications are often overgeneralized and overstated. Among the arguments they put forth are that nanoscale materials are frequently embedded in other materials as part of the manufacturing process; that some nanotechnology products, such as semiconductors, have nanoscale features but do not contain nanoscale particles; that nanotechnology materials may replace other materials that have significant and known risks; that some nanoscale particles tend to aggregate or agglomerate in the environment into larger particles that no longer have nanoscale dimensions; and that people are regularly exposed to nanoscale particles produced naturally and as incidental by-products of human activities.

Congressionally-mandated reviews of the NNI by the National Research Council (NRC) and the President's Council of Advisors on Science and Technology (PCAST) have concluded that additional research is required to make a rigorous risk assessment of nanoscale materials. In addition, the NRC warned that, until such information is available, precautionary measures should be taken to protect the health and safety of workers, the public, and the environment.

Nevertheless, most stakeholders agree that these concerns about the potential detrimental effects of nanoscale materials and devices—both real and perceived—must be addressed. Among the issues these stakeholders have identified are characterizing the toxicity of nanoscale materials; developing methods for assessing and managing the risks of these materials; and understanding how these materials move in, and interact with, the environment.

This report identifies the potential environmental, health, and safety opportunities and challenges of nanotechnology; explains the importance of addressing nanotechnology EHS concerns; identifies and discusses nanotechnology EHS issues; and summarizes options for Congressional action, including the nanotechnology EHS-related provisions of selected legislation. The report also includes two appendices. Appendix A provides an overview of selected nanotechnology EHS activities of federal regulatory agencies. Appendix B provides an overview of selected EHS-related international engagement efforts of NNI agencies.

For more information on nanotechnology and the NNI see, see CRS Report RL3440 1, *The National Nanotechnology Initiative: Overview, Reauthorization, and Appropriations Issues*, CRS Report RL34493, *Nan otechnology and U.S. Competitiveness: Issues and Options*, and CRS Report RL346 14, *Nanotechnology and Environmental, Health, and Safety: Issues for Consideration*, all by John F. Sargent. For additional information on nanotechnology-related regulatory challenges, see CRS Report RL34332, *Engineered Nanoscale Materials and Derivative Products: Regulatory Challenges*, by Linda-Jo Schierow.

OPPORTUNITIES AND CHALLENGES

Historically, many new technologies have delivered general societal benefits while presenting EHS challenges. For example, automobiles increased personal mobility and provided faster, less expensive transportation of goods, but soon became a leading cause of accidental deaths and injuries, as well as a source of emissions that can damage air quality and may contribute to global climate change. Similarly, genetically-modified (GM) plants have traits such as greater resistance to pests, pesticides, or cold temperatures that contribute to higher crop yields, while critics argue some GM foods contribute to food allergies and antibiotic resistance.[5]

Like other new technologies, nanotechnology offers potential economic and societal benefits, and presents potential EHS challenges as well. Nanotechnology advocates assert, however, that nanotechnology provides the opportunity to reduce or eliminate known risks by engineering around them. Proponents maintain that nanotechnology also offers the potential for significant EHS benefits, including:

- reducing energy consumption, pollution, and greenhouse gas emissions;
- cleaner, more efficient industrial processes;
- remediating environmental damage;
- curing, managing, or preventing deadly diseases; and
- offering new materials that protect against impacts, self-repair to prevent catastrophic failure, or change in ways that protect or aid soldiers on the battlefield.

For example, nanoscale materials show promise for detecting, preventing, and removing pollutants. According to the Environmental Protection Agency (EPA):

> nanoscale cerium oxide has been developed to decrease diesel engine emissions; iron nanoparticles can remove contaminants from soil and ground water; and nano-sized sensors hold promise for improved detection and tracking of contaminants.[6]

In the area of human health, scientists assert nanotechnology has the potential for improving disease diagnostics, sensing, monitoring, assessment, and treatment. In particular, the National Cancer Institute (NCI) views nanotechnology as likely to provide revolutionary tools to extend and improve lives. In July 2004, NCI launched a five-year, $145 million initiative focused on applying nanotechnology to the prevention, detection, and treatment of cancer and amelioration of its symptoms. At the initiative's launch, then-NCI Director Andrew von Eschenbach identified nanotechnology as a key component of the agency's strategy for ending death and suffering from cancer by 2015 (see text box, "Potential Nanotechnology Cancer Applications").[7]

Potential Nanotechnology Cancer Applications

The NCI Cancer Nanotechnology Plan asserts that nanotechnology can serve as an enabling technology for a variety of cancer-related applications:

- imaging agents and diagnostics that allow clinicians to detect cancer in its earliest, most easily treatable, pre-symptomatic stage;
- systems that provide real-time assessments of therapeutic and surgical efficacy;
- multifunctional, targeted devices capable of bypassing biological barriers to deliver therapeutic agents at high local concentrations directly to cancer cells and tissues that play a critical role in the growth and metastasis of cancer;
- agents capable of monitoring predictive molecular changes and preventing precancerous cells from becoming malignant;
- surveillance systems that detect mutations that may trigger the cancer process and genetic markers that indicate a predisposition for cancer;
- novel methods for managing the symptoms of cancer that adversely impact quality of life; and
- research tools that enable investigators to quickly identify new targets for clinical development and predict drug resistance.

Source: *Cancer Nanotechnology Plan: A Strategic Initiative to Transform Clinical Oncology and Basic Research Through the Directed Application of Nanotechnology*, National Cancer Institute, National Institutes of Health, Department of Health and Human Services, July 2004.

Some characteristics of nanoscale particles could produce both positive and negative consequences. According to E. Clayton Teague, director of the National Nanotechnology Coordination Office (NNCO),

> the unique properties of these [nanotechnology] materials are a double-edged sword: they can be tailored for beneficial properties, but also have unknown consequences, such as new toxicological and environmental effects.[8]

The following examples illustrate how the same nanotechnology material may be both potentially beneficial and potentially harmful:

- Nanoscale silver is highly effective as an antibacterial agent in wound dressings, clothing, and washing machines, but some have expressed concerns that widespread dispersion of nanoscale silver in the environment could kill microbes that are vital to waste water treatment plants and to ecosystems. Some beneficial bacteria, for example, break down organic matter, remove nitrogen from water, aid in animal digestion, protect against fungal infestations, and even aid some animals in defense against predators.[9]
- Some nanoscale particles may have the potential to penetrate the blood-brain barrier, a structure that protects the brain from harmful substances in the blood but also hinders the delivery of therapeutic agents. The characteristics of certain nanoscale materials may allow pharmaceuticals to be developed to purposefully and beneficially cross this barrier and deliver medicine directly to the brain to treat, for example, a brain tumor.[10] Some critics are concerned, however, that nanoscale particles might unintentionally pass through the blood-brain barrier causing harm to humans and animals.[11]
- Certain nanoscale materials are highly chemically reactive due to their high surface-to-volume ratio.[12] This is a property that might be positively exploited in catalysis, treatment of groundwater contamination, and site remediation. This property also is being explored for use in protective masks and clothing as a defense against chemical and biological agents. However, some research results indicate that the reactivity of some nanoparticles potentially can result in cell damage in animals.[13]
- Carbon nanotubes (CNTs) have potential uses in a wide range of applications (e.g., materials, batteries, memory devices, electronic displays, transparent conductors, sensors, medical imaging). However, some scientists have expressed concerns that some CNTs exhibit properties similar to asbestos fibers, and might become lodged in organs (e.g., lungs), harming humans and animals.[14]

EHS Concerns about Carbon Nanotubes and Other Fullerenes

Much of the public dialogue about potential risks associated with nanotechnology has focused on carbon nanotubes (CNTs) and other fullerenes (molecules formed entirely of carbon atoms in the form of a hollow sphere, ellipsoid, or tube) since they are currently being manufactured and are among the most promising nanomaterials. These concerns have been amplified by some research on the effects of CNTs on animals and on animal and human cells. For example, researchers have reported that carbon nanotubes inserted into the trachea of mice can cause lung tissue damage;[a] that buckyballs (spherical fullerenes) caused brain damage in fish;[b] and that buckyballs can accumulate within cells and potentially cause DNA damage.[c]

There are scientists who have argued that experiments indicating CNT/fullerene toxicity are not conclusive. They suggest that toxicity reported by researchers may have resulted from uncharacterized contaminants in the samples resulting from the synthesis, purification, and post-processing methods used in the manufacture of CNTs. Thus, they assert, the experiments could be measuring the toxicity of non-nanoscale materials and, therefore, unfairly indicting nanoscale materials. They also contend that such non-nanoscale contaminants, if identified as toxic, potentially could be eliminated or controlled in the manufacturing process. The issue of contaminants is often cited by advocates for improved standards, reference materials, sensors, instrumentation, and other technologies for the characterization of nanoscale materials.

Some experiments have produced results that indicate CNTs/fullerenes are non-toxic. Research on single-walled carbon nanotubes (SWCNTs) by the Institute of Toxicology and Genetics in Karlsruhe, Germany, reported that, in three of four different types of tests conducted, SWCNTs did not show toxicity. In the fourth test, which appeared to indicate SWCNT toxicity, the researchers concluded that the results were a "false positive" and explained how the SWCNTs interacted with the materials in the assay to produce a misleading result. These researchers concluded that this result points to the need for careful selection of assays and the need for the establishment of standards for toxicity testing of CNTs and other nanomaterials.[d]

Work at Rice University's Center for Biological and Environmental Nanotechnology conducted in 2005 found cell toxicity of CNTs to be low, and that it could be reduced further through simple chemical changes to the surface.[e] Earlier research demonstrated that similar surface modifications of buckyballs reduced their toxicity. Nanotechnology may offer the potential to engineer around known and potential hazards by changing the size, molecular construction, or other property of a nanoscale material to make it safe or less hazardous. Experts advise that the potential to do so will require a thorough understanding of the properties of the various nanoparticles and their effects on humans and other organisms.

[a] Lam, C.W., James, J.T., McCluskey, R., and Hunter, R.L. "Pulmonary toxicity of single-wall carbon nanotubes in mice 7 and 90 days after intratracheal instillation," September 2003. http://www.ncbi.nlm.nih.gov/sites/ entrez?cmd=Retrieve&db=PubMed&list_uids=14514958

[b] Oberdörster, Eva. "Manufactured Nanomaterials (Fullerenes, C60) Induce Oxidative Stress in the Brain of Juvenile Largemouth Bass," April 2004. http://www.pubmedcentral.nih. gov/articlerender.fcgi? tool=pmcentrez & artid=1247377

[c] Magrez, A., Kasas, S., Salicio, V., Pasquier, N., Seo, J.W., Celio, M., Catsicas, S., Schwaller, B., and Forro, L. "Cellular Toxicity of Carbon-Based Nanomaterials," *Nano Letters*, 6(6):1121-1125, American Chemical Society, May 2006. http://pubs.acs.org/cgi-bin/abstract.cgi/nalefd/ 2006/6/i06/ abs/ nl060 162e.html

[d] Wörle-Knirsch, J.M., Pulskamp, K., and Krug, H. F. "Oops They Did It Again! Carbon Nanotubes Hoax Scientists in Viability Assays," American Chemical Society, *Nano Letters*, Vol.6, April 2006.

[e] "Modifications render carbon nanotubes nontoxic," press release, Rice University, October 2005.

Importance of Addressing EHS Issues

Nanotechnology covers a wide swath of scientific fields, engineering disciplines, and technological applications. Sufficient knowledge has been developed about the useful properties of certain nanomaterials, how they can be manufactured, and how they can be applied in useful ways to enable commercial product development. In other areas of nanotechnology, fundamental research on nanoscale phenomena and processes is under way that may lead to greater understanding and beneficial applications in the years ahead. In general, however, nanotechnology is still an emerging field and there is a dearth of information about how nanoscale particles and devices might adversely affect human health, safety, and the environment. Accordingly, there is widespread agreement on the need for more research to better understand such implications.

In reviews of the NNI,[15] both the National Research Council and the President's Council of Advisors on Science and Technology (PCAST) concluded that assessment of potential nanotechnology EHS risks is not possible due to the absence of information and tools. According to the NRC,

> it is not yet possible to make a rigorous assessment of the level of risk posed by [engineered nanomaterials]. Further risk assessment protocols have to be developed, and more research is required to enable assessment of potential EHS risks from nanomaterials.[16]

Similarly, PCAST concluded that

> it is premature to rigorously assess the levels of risk posed by engineered nanomaterials. Adequate tools are being developed but are not yet in place.[17]

Leaders of the NNI have argued strongly that to achieve the economic, societal, and EHS benefits of nanotechnology the nation must concurrently address its potential adverse effects. According to then-Under Secretary of Commerce for Technology Phillip J. Bond, a leading Administration advocate for the NNI,

> Addressing societal and ethical issues is the right thing to do and the necessary thing to do. It is the right thing to do because as ethically responsible leaders we must ensure that technology advances human well-being and does not detract from it. It is the necessary thing to do because it is essential for speeding technology adoption, broadening the economic and societal benefits, and accelerating and increasing our return on investment.[18]

This is a view shared by many in the business community. A 2006 survey of business leaders in the field of nanotechnology indicated that nearly two-thirds believe that "the risks to the public, the workforce, and the environment due to exposure to nano particles are 'not known,'" and 97% believe that it is very important or somewhat important for the government to address potential health effects and environmental risks that may be associated with nanotechnology.[19]

The Project on Emerging Nanotechnologies (PEN) has warned that bad practices in nanotechnology research or production may result in a nanotechnology accident that would

> chill investment, galvanize public opposition, and generally lead to a lot of hand wringing on the part of governments who are betting large sums of money on the nanotech revolution.[20]

Successfully addressing EHS issues is seen as vital for those potentially exposed to nanoscale materials (e.g., consumers, researchers, manufacturing workers, the general public), businesses, and investors for a variety of reasons:

- protecting and improving human health, safety, and the environment;
- enabling accurate and efficient risk assessments, risk management, and cost-benefit trade-offs;
- ensuring public confidence in the safety of nanotechnology research, engineering, manufacturing, and use;
- preventing a problem in one application area of nanotechnology from having negative consequences for the use of nanotechnology in unrelated application areas due to public fears, legislative interventions, or an overly-broad regulatory response; and
- ensuring that society can enjoy the widespread economic and societal benefits that nanotechnology is believed by many to offer.

In addition, the U.S. regulatory environment for nanotechnology could be an enabler for innovation and contribute to a strong, sustainable economy by creating predictability, accurately assessing risks and benefits, and fostering the swift movement of safe products into the market. Such an environment is likely to favor nanotechnology-related investments and innovative activities in the United States by domestic and foreign stakeholders, as opposed to nations where such regulatory conditions do not exist.

Conversely, if the U.S. regulatory environment is not handled effectively (i.e., if it lacks predictability, if regulatory approaches do not accurately assess risks and benefits, or if approval processes are too long or expensive) it could prove a major impediment to innovation, economic growth, and job creation, as well as posing a potential threat to health, safety, and the environment. In such a regulatory environment, investment capital may be driven away from nanotechnology, potentially beneficial products may not be developed, safe products may be denied regulatory approval, or unsafe products may be allowed to enter the market.

Alternatively, nanotechnology investments, research, and production may be driven to other nations with preferable regulatory environments. On the one hand, such a regulatory system might be more desirable to investors and companies because it is more predictable, more efficient, and less costly. In such a case, the United States might miss out on nanotechnology's potential economic benefits. On the other hand, if other nations' regulatory systems are more attractive to investors and producers because those systems under-regulate or do not regulate at all, then nanotechnology research, development, and production could present increased EHS risks worldwide.

SELECTED ISSUES FOR CONSIDERATION

Given the widespread agreement that nanotechnology EHS concerns must be addressed, discourse on how best to do so has focused on three main issues:

- federal investment in EHS research;
- federal regulation; and
- international engagement.

These issues are closely interrelated. For example, reliable EHS research is required by regulatory bodies to determine whether and how to regulate nanotechnology products. Since all nations face the same fundamental health, safety, and environmental issues, international coordination on EHS research could help accelerate development of a common body of knowledge through the sharing of results and reduction in redundant research. This shared knowledge could, in turn, inform regulatory decision making and perhaps improve the consistency of regulations among nations. Regulations, standards, and enforcement might need to be coordinated worldwide to protect workers and consumers as intermediate and final products are frequently produced along global supply chains and sold in industrial and commercial markets around the world. In addition, one nation's policies governing nanotechnology production, use, and disposal may have implications for nearby nations and, perhaps, for all nations.

Federal Investment in EHS Research

Current Funding Level

There is not a single, centralized source of EHS research funds that is allocated to individual agencies. Agency nanotechnology budgets are developed internally as part of each agency's overall budget development process. These budgets are subjected to review, revision, and approval by the Office of Management and Budget (OMB) and become part of the President's annual budget submission to Congress. The NNI budget—and the EHS component—is then calculated by aggregating the nanotechnology components of the appropriations provided by Congress to each federal agency. While there is some coordination of EHS-research budget requests through the Nanotechnology Environmental and Health Implications (NEHI) working group[21] and in OMB's budget development process, the decision process that establishes overall funding for nanotechnology EHS research is highly decentralized.

In FY2008, NNI funding for EHS implications research[22] was $58.6 illion, approximately 3.9% of the total NNI budget of $1.49 billion. This represented an increase over the FY2007 EHS research level of $48.3 million (3.4% of the total NNI budget), and the FY2006 level of $37.7 million (2.8%), both in dollars and in share of total NNI funding. President Bush requested $76.4 million (5.0%) for EHS research in FY2009. NNI EHS research funding for FY2006 through FY2008, and the request for FY2009, is provided in **Table 1.**

Table 1. NNI Environmental, Health, and Safety Research Funding, FY2006-2008, FY2009 Request

	EHS research, in current dollars	EHS research's share of total NNI budget
FY2006 (actual)	$ 37.7 million	2.8%
FY2007 (actual)	48.3 million	3.4%
FY2008 (estimated)	58.6 million	3.9%
FY2009	76.4 million	5.0%

Sources: "The National Nanotechnology Initiative: Research and Development Leading to a Revolution in Technology and Industry, Supplement to the President's FY2008 Budget," NSET Subcommittee, NSTC, OSTP, The White House, July 2007; "National Nanotechnology Initiative: FY2009 Budget and Highlights," NSET Subcommittee, NSTC, The White House, February 2008. http://www.nano.gov/NNI_ FY09_budget_ summary.pdf

NNI officials assert that the initiative also conducts EHS research as a part of its other research activities, but that these EHS investments are not easily quantified and thus are not reflected in the NNI's reported figure for EHS funding. PCAST agreed with this assertion in its 2008 assessment, arguing that

> In many instances, nanotechnology EHS research cannot be separated from the particular application(s) research and from the context for which a specific nanomaterial is intended. Such division is unproductive and neglects the whole benefit of research. Consequently, [PCAST] expects that a substantial fraction of nanotechnology research related to EHS will continue to take place under the auspices of agencies that fund applications R&D and may not be uniquely or exclusively identified as nanotechnology EHS research.... Furthermore, detailed reporting on the degree of relevance to EHS of such research is not necessarily critical to (and may actual hinder) overall prioritization and coordination.[23]

In 2007, OMB issued a one-time request to all NNI research agencies to report funding data on research related to the five categories identified in the NSET document, *Prioritization of Environmental, Health, and Safety Research Needs for Engineered Nanoscale Materials*.[24] Totals for EHS implications research spending identified in each of the five categories is shown below in **Table 2**. Preliminary analysis of this data by the NEHI working group indicated that $67 million was spent on EHS research in FY2006, in contrast to the reported figure of $37.7.

Table 2. FY2006 NNI Funding for EHS Research *by Research Needs Categories*

Category	Estimated Funding
Instrumentation, Metrology, and Analytical Methods	$27 million
Nanomaterials and Human Health	$24 million
Nanomaterials and the Environment	$13 million
Health and Environmental Exposure Assessment	$ 1 million
Risk Management Methods	$ 3 million
TOTAL	**$67 million**

Source: Teague, E. Clayton, director, National Nanotechnology Coordination Office. Testimony before the Subcommittee on Research and Science Education, Committee on Science and Technology, U.S. House of Representatives. Hearing on "Research on Environmental and Safety Impacts of Nanotechnology: Current Status of Plan ning and Implementation under the National Nanotechnology Initiative." 110th Cong., 1st Sess., October 31, 2007. Note: Numbers may not add due to rounding.

Critics (as detailed in the following section) assert that the current level of federal nanotechnology EHS research is too low and represents too small a share of the overall NNI budget. These critics argue that the current allocation of NNI funding may produce a flood of products for which there is inadequate information to assess and manage their EHS risks.

However, executive branch officials stress that the United States leads the world in EHS funding and, by inference, that the current funding level is adequate. White House Office of Science and Technology Policy (OSTP) director John Marburger asserted that the United States

> leads the world not only in spending for nanotechnology development, but also, by an even larger margin, in its investment in research to understand the potential health and safety issues.[25]

Similarly, NNCO director E. Clayton Teague asserted U.S. leadership in nanotechnology EHS research:

> During fiscal years 2005 through 2008, it is estimated that NNI agencies will have invested nearly $180 million in research whose primary purpose is to address the EHS implications of nanomaterials. With these investments, the United States leads all other countries by a wide margin in support of such research.[26]

Dr. Teague maintains that EHS research has been a top priority of the Administration and the NNI, citing, as an example, the annual R&D budget guidance memorandum sent by the directors of OMB and OSTP to departments and agencies. This memorandum identifies Administration priorities and is intended to help guide agency budget development for the following fiscal year. The OMB/OSTP memorandum to guide FY2006 agency budget development stated that

> In order to ensure that nanotechnology research leads to the responsible development of beneficial applications, agencies also should support research on the various societal implications of the nascent technology. In particular, agencies should place a high priority on research on human health and environmental issues related to nanotechnology and develop, where applicable, cross-agency approaches to the funding and execution of this research.[27]

The OMB/OSTP memorandum has included similar language in each succeeding year.

In their reviews of the NNI, both the NRC and PCAST concluded that federal EHS research funding should be expanded. According to the NRC assessment,

> To help ensure the responsible development of nanotechnology ... research on the environmental, health, and safety effects of nanotechnology [should] be expanded.[28]

PCAST acknowledged potential EHS risks in its first review of the NNI but found the federal government was "directing appropriate attention" and "adequate resources" to EHS research. In its second assessment, PCAST termed the current federal investment level in EHS "appropriate," but added that

> expanded EHS research, broad-based protocol development, and particularly standardization are necessary.... the funding level for EHS [should] continue to grow consistent with the needs identified in the NNI research strategy for nanotechnology EHS as well as the available capacity for quality research.[29]

Alternative Approaches

Various alternatives have been suggested for addressing the perceived shortcoming in EHS funding. One recommendation is requiring a fixed percentage of the NNI's total funding be devoted to EHS research. A figure of 10% has been proposed for this purpose by organizations such as the NanoBusiness Alliance and the Project on Emerging Nanotechnologies. If this proposal had been in effect in FY2008, the NNI would have been required to spend $149 million on EHS research, more than twice as much as the NSET-reported level of $58.6 million. In testimony before the House Committee on Science and Technology, Sean Murdock, executive director of the NanoBusiness Alliance, agreed with the level of funding represented by the 10% figure but argued the need for cross-agency flexibility in achieving it:

> The NanoBusiness Alliance believes that environmental, health, and safety research should be fully funded and based on a clear, carefully-constructed research strategy. While we believe that 10 percent of the total funding for nanotechnology research and development is a reasonable estimate of the resources that will be required to execute the strategic plan,

> we also believe that actual resource levels should be driven by the strategic plan as they will vary significantly across agencies.[30]

Others have suggested a different approach, proposing fixed dollar amounts or minimum levels. For example, the Environmental Defense Fund has called for $100 million or more in federal nanotechnology EHS research funding.[31]

In its 2008 assessment, PCAST disagreed with both approaches:

> growing research in nanotechnology EHS must be strategic, guided by ... a comprehensive set of scientifically determined priorities and needs rather than arbitrary percentages or funding figures.[32]

By establishing a 10 percent requirement (or setting a figure of $100 million for total EHS funding), the United States could greatly accelerate the growth in EHS research spending. In testimony before Congress in 2007, PCAST co-chair Floyd Kvamme warned against such a rapid increase:

> In general, increasing funding too rapidly does not lead to equivalent increases in high quality research. It is crucial to note that EHS research also depends on advances in non-EHS areas, such as instrumentation development and basic research on nanomaterials.[33]

Some non-governmental organizations (NGOs) have advocated for a more restrained approach to nanotechnology research and development. They assert that the federal government is pushing ahead too quickly in developing nanotechnology and encouraging its commercialization and use without sufficient knowledge and understanding of EHS implications and how they might be mitigated.[34] They argue that the very characteristics that make nanotechnology promising also present significant potential risks to human health and safety and the environment. Some of these groups argue for application of the "precautionary principle,"[35] which holds that regulatory action may be required to control potentially hazardous substances even before a causal link has been established by scientific evidence.[36] In 2006, Friends of the Earth warned that

> The early warning signs surrounding nanotoxicity are serious and warrant a precautionary approach to the commercialization of all products containing nanomaterials ... . there should be a moratorium on the further commercial release of sunscreens, cosmetics and personal care products that

> contain engineered nanomaterials, and the withdrawal of such products currently on the market, until adequate public, peer-reviewed safety studies have been completed, and adequate regulations have been put in place ... [37]

The Action Group on Erosion, Technology, and Concentration (ETC Group) has called for a moratorium on the conduct of nanotechnology R&D and use of commercial products incorporating man-made nanoparticles:

> Given the concerns raised over nanoparticle contamination in living organisms, Heads of State ... should declare an immediate moratorium on commercial production of new nanomaterials and launch a transparent global process for evaluating the socio-economic, health and environmental implications of the technology.[38]

In 2003, the ETC Group expanded the breadth of its proposed moratorium:

> In the absence of toxicology studies, ETC Group believes that governments must also urgently consider extending the moratorium to products that place consumers in direct contact with synthetic nanoparticles through their skin, lungs or digestive systems.[39]

In contrast to these views, a report prepared by the NSET Subcommittee concluded that conducting EHS research in parallel with the development of nanomaterials and their applications will help to ensure the full, safe, and responsible realization of the promise of nanotechnology.[40]

In 2003, then-Under Secretary of Commerce for Technology Phillip J. Bond addressed calls for a moratorium or slowdown in nanotechnology R&D, casting the issue in ethical terms:

> Those who would have us stop in our tracks argue that it is the only ethical choice. I disagree. In fact, I believe a halt, or even a slowdown, would be the most unethical of choices.... Given the promise of nanotechnology, how can our attempt to harness its power at the earliest opportunity—to alleviate so many of our earthly ills—be anything other than ethical? Conversely, how can a choice not to attempt to harness its power be anything other than unethical?[41]

Management of Federal EHS Research

In order to manage the Federal EHS portfolio, policymakers will need to establish research priorities. In this regard, the NRC recommended that

> Assessing the effects of engineered nanomaterials on public health and the environment requires that the research conducted be well defined and reproducible and that effective methods be developed and applied to (1) estimate the exposure of humans, wildlife, and other ecological receptors to source material; (2) assess effects on human health and ecosystems of both occupational and environmental exposure; and (3) characterize, assess, and manage the risks associated with exposure.[42]

In 2005, PCAST concluded that EHS research should give highest priority to workplace exposure. PCAST noted

> the greatest likelihood of exposure to nanomaterials is during manufacture, and therefore [we] agree with the prioritization of research on potential hazards from workplace exposure.[43]

Several years later, in its 2008 assessment, PCAST reiterated this point stating, "the greatest risk of exposure to nanomaterials at present is to workers who manufacture or handle such material," but also acknowledged a broader range of risks:

> environmental, health, and safety risks in a wide range of settings must be identified and the necessary research performed so that real risks can be appropriately addressed.[44]

Some stakeholders have asserted that a comprehensive approach to federal EHS research has been hampered by the lack of an NNI roadmap for these efforts.[45] In general, these stakeholders seek a multi-year roadmap with specific milestones, metrics, and funding levels. Such a roadmap, they assert, would contribute to a more coordinated approach among agencies and between the executive branch and Congress on the magnitude, timing, prioritization, and management of federal EHS research.

NNI officials argue that the NSET Subcommittee, the coordinating body for the NNI, has developed an EHS research strategy and articulated it in three reports (see text box, "NNI EHS-focused Reports"), though they acknowledge that these documents do not constitute a roadmap. At an October 2007 hearing of the House Subcommittee on Research and Education,[46] some Members of

Congress expressed concerns about the time required by the National Nanotechnology Coordination Office to produce a prioritized, detailed implementation plan for NNI EHS research. While acknowledging the challenges faced by the NNCO in developing consensus among the 25 NNI agencies, some Members suggested that these challenges were emblematic of the need for a more top-down approach to EHS research.

NNI EHS-FOCUSED REPORTS

Environmental Health and Safety Research Needs for Engineered Nanoscale Materials, published in September 2006, identified the research and information needed to enable sound risk assessment and risk management decision making with respect to nanoscale materials and products that incorporate them.

Prioritization of Environmental, Health, and Safety Research Needs for Engineered Nanoscale Materials, published in August 2007, identified five broad categories of EHS research and information needs, and five specific research areas in each category.

The National Nanotechnology Initiative: Strategy for Nanotechnology-related Environmental, Health, and Safety Research, published in February 2008, defined the NNI's strategy for addressing priority research on EHS aspects of nanomaterials. The document reviewed current agency research using the taxonomy developed in the second report; identified research gaps; and articulated a framework for prioritizing research, implementing the strategy, and coordinating agency efforts.

Opposition to an EHS roadmap stems primarily from doubts of the practicality and efficacy of such an approach. Some argue that it is unlikely that OMB would commit to a multi-year, multi-agency roadmap accompanied by specific funding levels. Such an approach would depart from the current executive branch annual budget development process and reduce OMB's flexibility in future years. In addition, agencies often have to respond to new requirements based on emergent circumstances, Congressional direction, or other factors. Agency funding is often redirected from planned efforts to new, often imminent, priorities. The need for such redirection of funding could impede the achievement of roadmap milestones and metrics or, conversely, impede the movement of funding to new priorities.

To overcome the obstacles associated with the development of a roadmap by the agencies, some have suggested the National Academies produce such a roadmap. Some assert that this approach worked well with respect to the development of a federal research roadmap to reduce EHS uncertainties associated with airborne particulate matter. Others argue that the particulate matter effort focused only a narrow field and covered research conducted by only a single agency (EPA); in contrast, nanotechnology spans a broad range of materials and applications across many fields, and requires EHS research efforts by several agencies.

In February 2007, 19 environmental and business organizations, large and small companies, and research organizations signed a letter to the Senate Appropriations Subcommittee on Interior, Environment, and Related Agencies requesting $1 million be appropriated for the development of a federal roadmap and research strategy. The letter recommended that this work be done by the National Institute of Environmental Health Sciences (NIEHS).[47]

The Senate Appropriations Committee report (S.Rept. 110-91) accompanying the Department of the Interior, Environment, and Related Agencies Appropriations Act, 2008[48] urged the Environmental Protection Agency (EPA) to

> contract or enter into a cooperative agreement with the National Academy of Sciences' Board on Environmental Studies and Toxicology within 90 days of enactment to develop and monitor implementation of a comprehensive, prioritized research roadmap for all Federal agencies on environmental, health and safety issues for nanotechnology.[49]

A COOPERATIVE APPROACH TO ADDRESSING EHS CONCERNS

Some organizations have taken a cooperative approach to promote EHS research. For example, the Environmental Defense Fund, an environmental advocacy group, partnered with the American Chemistry Council, a trade group, to issue a Joint Statement of Principles in June 2005 that recognizes the "significant societal and sustainable development benefits" expected from nanotechnology, while calling for a multi-stakeholder dialogue to achieve the timely development of nanomaterials "in a way that minimizes potential risks to human health and the environment." The statement also called for increased federal investments in EHS research and development of an international effort to standardize testing protocols, hazard and exposure assessment approaches, and nomenclature and terminology ... to maximize resources and minimize inconsistent regulation of nanomaterials.[a]

There is general agreement among stakeholders that these activities can contribute to creating an environment where research results can be reliably shared and compared, to protecting human health and safety, and to creating a common language about nanotechnology that increases clarity in the sharing of ideas and information. However international standardization efforts are often time- and resource-consuming, and can divert resources from more pressing needs. In addition, such efforts can be used by nations and other organizations for competitive advantage (e.g., by securing the adoption of a favorable standard, slowing others' progress).

In June 2007, the Environmental Defense Fund and DuPont issued a Nano Risk Framework "to assist with the responsible development and use of nanotechnology and to help inform global dialogue on its potential risks."[b] The framework is a six-step process to identify, address, and manage potential risks: (1) describe the material and the intended application; (2) profile the material's lifecycle in the application; (3) evaluate associated risks; (4) assess risk management options; (5) decide on and document actions; and, (6) regularly review new information and adapt actions accordingly.[c]

[a] *Environmental Defense and American Chemistry Council Nanotechnology Panel: Joint Statement of Principles*, Comments on EPA's Notice of a Public Meeting on Nanoscale Materials, June 23, 2005.

[b] "DuPont and Environmental Defense Launch Comprehensive Tool for Evaluating and Addressing Potential Risks of Nanoscale Materials," press release, E. I. du Pont de Nemours and Company, June 21, 2007.

http://vocuspr.vocus.com/VocusPR30/Newsroom/Query.aspx?SiteName=Dupont New&Entity=PRAsset&SF_PRAsset_PRAssetID_EQ=106677&XSL=PressRelease&Cac he=False

[c] *Nanorisk Framework*, Environmental Defense-DuPont Nano Partnership, June 2007.

The process used to develop research priorities and the federal EHS budget has also raised management concerns. As discussed earlier, the federal nanotechnology EHS research portfolio results from research funding requests made by individual agencies pursuing their missions and by decisions made in the Congressional appropriations process. Informal research coordination among EHS funding agencies occurs through the NEHI working group and more formally through the OMB budget development process. Some proponents for an integrated federal EHS research effort have called for a more top-down approach. The Woodrow Wilson Center's Project on Emerging Nanotechnologies (PEN) has been a leading advocate on this issue. PEN's chief science advisor, Andrew Maynard, asserted that

> to realize nanotechnology's benefits ... the federal government needs a master plan for identifying and reducing potential risks. This plan should include a top-down risk research strategy, dedicated and sufficient funding to do the job, and the mechanisms to ensure that resources are used effectively.[50]

PEN has recommended increasing the authorities of the NEHI working group to empower it to develop and implement the top-down research plan, a minimum of $100 million over two years to fund the research, and a full-time director to support the NEHI working group.

Responding to the PEN recommendation, E. Clayton Teague, director of the NNCO, testified before Congress that there was a consensus among NNI agencies that a centralized office with budgetary authority to oversee the NNI's EHS research program would have significant detrimental effects. According to Dr. Teague,

> No one agency or centralized organization would have the breadth of scientific expertise and knowledge of regulatory authorities and needs currently represented by the 20 agencies participating in the NEHI working group.
>
> Creation of a new central authority would undermine the existing successful interagency coordination.
>
> Moving the management of all nanotechnology EHS research into a single office would likely decouple such research from related efforts within NNI agencies and from the knowledge base in the agencies that is currently networked into the NNI's EHS research effort.
>
> Creating a separate office would, on the one hand, give mission agencies a disincentive for doing nanotechnology-related EHS research. They would reasonably assume that another agency is responsible, and they therefore could redirect their limited resources to address other priorities. A likely result could be that the level of research would actually decrease. Conversely, creating a separate office could lead to duplicative work being funded, thereby wasting tax dollars and not optimizing progress.[51]

Project on Emerging Nanotechnologies Recommendations

The Project on Emerging Nanotechnologies (PEN), a joint venture of the congressionally-chartered Woodrow Wilson Center for International Scholars and the Pew Charitable Trusts, has produced inventories of both nanotechnology-based products and government-funded EHS research. PEN has asserted the need for more EHS research, more aggressive oversight, and a more centralized federal government approach to funding EHS research.

In addition, PEN contends that the increasing complexity of systems incorporating nanoparticles with multiple functions will make the behaviors more complex and difficult to predict. To minimize the likelihood of a nanotechnology accident, PEN made the following recommendations:

- Creating a Nano Safety Reporting System where people working with nanotechnology can anonymously report safety issues and concerns. PEN states that the information gleaned from this system could be used to inform the design of educational materials, better structure technical assistance programs, and provide an early indicator of emerging safety issues.
- Creating technologies that provide an early-warning system to allow for risk to be assessed early in research efforts. Such a technology might enable low-cost, fast-screening for novel properties that would allow for risk assessment integrated and concurrent with the R&D process.
- Pushing information out to small businesses, start-ups, and laboratories that, due to their size and resources, are unlikely to be able to devote significant resources to EHS issues. PEN states that existing assistance programs could be used to deliver this information, as well as the development of peer-to-peer mentoring programs within industrial supply chains.
- Application of lessons learned in other technology areas to make nanotechnology more inherently safe, using strategies such as multiple levels of protection, learning from failures, not oversimplifying the complex, awareness of operations, and building in resilience to prevent cascading of errors.

Source: Rejeski, David, director, Project on Emerging Nanotechnologies. "Nanotech Safety 101 or How to Avoid the Next Little Accident," paper, Workshop on Disaster Prevention, Harvard University, April 27, 2006.

Dr. Maynard counters that "it should be possible to develop a functional structure that enables agencies to work within a broader plan." According to Maynard, while a centralized office is not necessary,

> top-down leadership with authority and the ability to ensure resources get to where they are needed is necessary.... [Such] leadership does not take away from agencies' expertise and missions, but rather empowers agencies to do the best they can, while coordinating and partnering as effectively as possible with each other.[52]

Federal Regulation

Some have raised concerns about whether current laws, regulations, and authorities are adequate to protect human health, safety, and the environment from potential adverse implications of nanotechnology. Several factors may affect the ability of the regulatory system to keep pace with advances in technology, both broadly and specifically with respect to nanotechnology.

Broadly, market forces have increased the pace of global innovation, challenging institutions' ability to identify and cope with the societal implications of rapid change. Speed-to-market has become a driving factor in competition for many industries as a result of the entry of new and nimble competitors in the global marketplace, increased public and private investments in R&D, global models of innovation, increased flows of scientific and technical knowledge, and greater numbers of scientists and engineers around the world. In addition, growing global markets enable companies to recoup their investments faster and enable earlier investments in subsequent generations of technology, further accelerating the pace of innovation. The increased pace, scope, and complexity of technological innovation may pose challenges to the existing regulatory system. While these factors may affect a broad range of technologies, nanotechnology may be especially affected due to the rapid growth in public and private R&D investments in the field since the year 2000 and the potential for nanomaterials to be used in a wide array of products.

Nanotechnology also may pose unique challenges to the regulatory system. For example, historically, regulatory agencies have defined a chemical by its chemical composition, usually without regard to its particle size. In contrast, the essence of nanotechnology is that a material may exhibit different properties at the nanoscale than it does at a bulk, molecular, or atomic scale.

(See text box, "Unique Properties Emerge at the Nanoscale.") Accordingly, questions are being raised by representatives of the scientific, advocacy, and regulatory communities about how an EHS research portfolio might be structured when particle size may affect a material's properties, whether it may be necessary to incorporate particle size into regulatory regimes, and how this might be accomplished given the vast spectrum of particle sizes that might affect the characteristics of a particular material.

UNIQUE PROPERTIES EMERGE AT THE NANOSCALE

Scientists have discovered that elements and materials with the same chemistry can exhibit fundamentally different properties at the nanoscale. For example, platinum, which exhibits no magnetism in its bulk form, shows significant magnetic properties in nanoscale clusters of 13 atoms. The optical properties of gold also can change with particle size. At 10 nanometers, gold particles absorb green light and appear red, not gold.

Not only can nanoscale particles differ in properties from bulk material with the same chemical composition, they may also differ from other nanoscale materials with the same chemical composition. For example, the melting point of an element—which was believed to be constant regardless of the element's particle size—can change with particle size. Nanotechnology research has demonstrated that the melting temperature of gold decreases when the particle's radius drops below 10 nanometers (from a melting temperature of approximately 1,000°C at 10 nanometers to approximately 500°C at 2 nanometers).

Source: Roduner, Emil. "Nanoscopic Materials: Size- Dependent Phenomena," University of Stuttgart, Germany, August 2006.

Some argue that EHS concerns about nanotechnology products can be handled under existing laws and regulations, while others see legal obstacles to adequate EHS regulation. In both of its assessments of the NNI, PCAST concluded that existing regulatory authorities were adequate for the current activities; that appropriate regulatory mechanisms should be used to address instances of harmful human or environmental effects of nanotechnology; and that new regulatory policies related to nanotechnology should be rational, science-based, and consistent across the federal government. Similarly, Sean Murdock, executive director of the NanoBusiness Alliance, asserted that

> The apparatus for effective nanotechnology regulation is largely in place through various statutes and agencies, but it lacks data and resources. To enable these agencies and for the nanotech regulation effort to succeed we must increase the level of funding available to them for nanotech environmental, health and safety research; coordinate efforts between agencies; establish metrics and standards that can be used to characterize nanomaterials; conduct ongoing research; and more.[53]

Others believe that new laws and regulations, or modifications to existing ones, may be required. J. Clarence Davies, senior advisor to the Project on Emerging Nanotechnologies and former EPA Assistant Administrator for Policy, Planning, and Evaluation argued that

> Nanotechnology is difficult to address using existing regulations. There are a number of existing laws—notably the Toxic Substances Control Act; the Occupational Safety and Health Act; the Food, Drug and Cosmetic Act; and the major environmental laws (Clean Air Act, Clean Water Act, and Resource Conservation and Recovery Act)—that provide some legal basis for reviewing and regulating [nanotechnology] materials. However, all of these laws either suffer from major shortcomings of legal authority, or from a gross lack of resources, or both. They provide a very weak basis for identifying and protecting the public from potential risk, especially as nanotechnologies become more complex in structure and function and the applications become more diverse.
>
> A new law may be required to manage potential risks of nanotechnology. The law would require manufacturers to submit a sustainability plan which would show that the product will not present an unacceptable risk.[54]

Davies further asserts that new mechanisms and institutional capabilities—including research programs, tax breaks, acquisition programs, and regulatory incentives—are needed to encourage beneficial applications of nanotechnology.

In developing the regulatory structure, some in the business and financial communities assert that stability and predictability are key characteristics for attracting investment and spurring commercial applications. According to Matthew Nordan, vice president of Lux Research, the ambiguity surrounding environmental, health, and safety regulation of nanoparticles is hampering commercialization. Firms do not want to play a game whose rules may change at any time.... That doesn't mean they want more regulations or more onerous regulations. They're just looking for a roadmap on how federal agencies such as the EPA or OSHA [Occupational Safety and Health Administration] plan to approach nanoparticles.[55]

Some tension exists between the goals of promoting the development of nanotechnology, ensuring the global competitive position of the United States, addressing potential EHS implications of nanotechnology, and coping with the unique challenges nanotechnology poses to the current regulatory regime. To prevent health and safety concerns from becoming an impediment to innovation, some suggest that health and safety research and regulation must be done near-concurrently with product development, keeping pace with the speed of innovation. Alternatively, others argue that the potential health, safety, and environmental implications are either unknown or of such significance that EHS research and regulation must precede nanotechnology development and commercialization. "By the time monitoring catches up to commerce the damage will already have been done," asserted Ian Illuminato, health and environment campaigner for Friends of the Earth.[56] AFL-CIO industrial hygienist Bill Kojola warned that

> Even though potential health hazards stemming from exposure have been clearly identified, there are no mandatory workplace measures that require exposures to be assessed, workers to be trained, or control measures to be implemented. [Nanotechnology] should not be rushed to market until these failings are corrected and workers assured of their safety.[57]

The National Research Council assessment of the NNI acknowledged the need for additional reproducible, well-characterized EHS data to inform risk-based guidelines and best practices and warned that until such information is available precautionary measures should be taken to protect the health and safety of workers, the public, and the environment.[58]

In its 2008 assessment of the NNI, PCAST asserted that risk research must not be considered in isolation, but rather in the context of the overall risks and benefits of a particular material or technology. This perspective is shared by many industry advocates who argue that regulatory decisions must balance the potential risks associated with a nanotechnology product against the benefits it delivers and the risk it displaces. Further, they maintain that nanotechnology products should not be held to a higher standard than non-nanotechnology products. PCAST also noted that manufacturers and sellers of nanotechnology products had responsibilities for ensuring workplace and product safety, and asserted that the NNI has a vital role in supporting federal regulatory agencies by providing them with EHS research results.

A description of selected nanotechnology EHS activities of federal regulatory agencies is provided in **Appendix A**.

International Engagement

International engagement on EHS issues is believed by many to be important to the responsible development and successful commercialization of nanotechnology. NNI officials assert that the United States has played a central role in convening international efforts to address EHS concerns. In its 2008 assessment, PCAST encouraged the NNI to coordinate its efforts with other nations to avoid duplication and to leverage investments, characterizing such work as "noncompetitive."[59]

Federal agencies have engaged internationally (e.g., with agencies of other nations, international organizations, standards organizations) across a wide range of nanotechnology-related areas, including standards, nomenclature, and EHS research. Appendix B provides an overview of selected international engagement efforts of NNI agencies related to environmental, health, and safety issues.

Advocates for international engagement assert a variety of potential benefits. For example, transparency and/or harmonization of standards and regulations may contribute to assurance of global supply chains and market confidence in nanotechnology products. Increased globalization of production and markets means that companies and consumers around the world are increasingly part of a common network. Manufacturers of final products generally rely on inputs from multiple suppliers in their global supply chains. The reliability of a final product often depends on the reliability of inputs, such as materials or components. Transparent and common standards and regulations may help to ensure the integrity of supply chains and final products. While this is an issue for a variety of non-nanotechnology products (e.g., the recent discovery of lead-tainted toys and other products imported from China), nanotechnology may present a unique challenge in that at least some nanoscale particles can be incorporated into materials and products in ways that cannot be easily detected or detected at all. Thus, producers and the consumers they serve must rely, in large measure, on standards and regulatory systems to ensure that nano scale materials are properly produced and represented throughout the supply chain. In the absence of such standards and regulatory systems, producers may not be able to rely on inputs or may incur additional costs for testing and verification; substandard inputs may be incorporated in final products making them underperform or unsafe, and possibly resulting in loss of market confidence and/or potential litigation; or nanotechnology materials may be incorporated without disclosure.

Internationally agreed upon standards could also contribute to greater comparability of research results, improving understanding of EHS-related aspects of nanotechnology, and promoting regulations that help protect human health and the environment. Common standards and nomenclature also may contribute to more effective global collaboration in nanoscale science, engineering, and technology R&D, accelerating the realization of nanotechnology's economic and societal potential.

Global engagement may help to establish a common environment for the development and production of nanotechnology products and to promote access to global markets. In the absence of such an environment, some nations may seek to attract investments in their markets by adopting lower environmental, health, and safety standards and regulations.

Finally, while much remains unknown about the transport and fate of nanoscale materials released into the environment, it is possible that countries and populations other than those where research and production activities take place may be affected. Efforts to promote the adoption of best practices in nanotechnology research, production, use, disposal, and recycling may protect human health and the environment worldwide.

International engagement on EHS research may pose problems, including the time, cost, difficulty, and alleged ineffectiveness of such collaborations. For example, while some advocates assert the need for swift action in advancing EHS research, international engagements often entail slow processes. Also, given the strong U.S. position in nanotechnology, broadly, and in nanotechnology EHS research, specifically, some may argue that other countries have little to contribute, that such efforts tax limited federal EHS financial and human resources, and that such diffusion of resources may slow overall EHS progress. Others might assert that international engagement efforts focused explicitly on nanotechnology are unnecessary given the wide variety of existing mechanisms and pathways for sharing academic research and environmental, health, and safety information across national borders.

Some may oppose international engagement efforts because they lack faith in the goodwill of some participating parties due to the potentially strong national interests at stake (e.g., military applications, economic growth, job creation). In 2003, then-Under Secretary of Commerce for Technology Phillip

J. Bond questioned whether global calls for a slowdown in nanotechnology R&D to address environmental, health, and safety concerns are intended to allow other nations to close the nanotechnology leadership gap with the United States:

> I wonder very often if there are really calls for a slow-down so that other governments and countries might catch up.[60]

Others assert that the research required to understand and address EHS implications may be closely linked to applications-related R&D to create nanotechnology materials, products, or processes. In such cases, companies and countries may be reluctant to reveal EHS concerns and efforts, to cooperate in EHS research, or to share results as such actions may reveal competitive strategies, provide information others might use to compete against them (e.g., insights into promising materials or manufacturing processes), or result in unwanted scrutiny by regulators.

Concluding Observations

Advocates and critics agree that potential environmental, health, and safety implications of nanotechnology must be addressed if the full economic and societal benefits of nanotechnology are to be achieved. There is also general agreement that the current body of knowledge of how nanoscale materials might affect humans and the environment is insufficient to assess, address, and manage the potential risks. While there is agreement on the need for more EHS research, there are differing views on the level of funding required, how it should be managed, and related issues.

Congress is currently considering legislation, H.R. 554, that would reauthorize and amend the 21st Century Nanotechnology Research and Development Act, the appropriations bills that fund the NNI agencies' nanotechnology EHS research, and H.R. 820, the Nanotechnology Advancement and New Opportunities Act. Congress may use these opportunities to further address nanotechnology EHS implications issues, including: How much should the federal government appropriate for EHS research? How can the federal EHS research investment be better accounted for? How should the research be prioritized? Should the research be more centrally managed? How can EHS research results and best practices be shared more broadly? Can voluntary programs effectively provide needed information about industrial nanotechnology production activities? How can efforts to develop common nomenclature and standards be improved? Are existing laws, regulations, guidelines, and regulatory structures adequate? Is there sufficient coordination among federal regulatory agencies? What types of international engagement on nanotechnology research and regulatory issues could best foster responsible development of nanotechnology?

NANOTECHNOLOGY EHS-RELATED LEGISLATION IN THE 111TH CONGRESS

Two bills introduced in the 111th Congress contain provisions that seek to address nanotechnology EHS concerns. The following section summarizes selected EHS-related provisions of these bills.

H.R. 554—National Nanotechnology Initiative Amendments Act of 2009

H.R. 554, the National Nanotechnology Initiative Amendments Act of 2008, was introduced on January 15, 2009, and referred to the House Committee on Science and Technology. This act would revise the 21st Century Nanotechnology Research and Development Act in a variety of ways, several of which specifically address nanotechnology EHS concerns. The legislation:

- directs the National Nanotechnology Coordination Office to develop and maintain a public database of NNI EHS projects, including the agency funding source and funding history;
- requires the National Nanotechnology Advisory Panel (NNAP) to be established as a "distinct entity" (the NNAP's functions are currently performed by the President's Council of Advisors on Science and Technology), and requires the establishment of a subpanel to assess whether societal, ethical, legal, environmental, and workforce concerns are adequately addressed by the NNI;
- directs that the National Research Council, as part of its triennial review of the NNI, evaluate the adequacy of the NNI's efforts to address ethical, legal, environmental, human health, and other appropriate societal concerns;

- requires the designation of an associate director of the White House Office of Science and Technology Policy to serve as Coordinator for Societal Dimensions of Nanotechnology with responsibility for developing an annual research plan for federal nanotechnology EHS activities, monitoring and encouraging agency EHS efforts, and for encouraging agencies to engage in public-private partnerships to support EHS research;
- requires certain interdisciplinary research centers supported under the NNI to include EHS research to develop methods for developing environmentally benign nanoscale products and processes, to foster the transfer of research results to industry, and to provide interdisciplinary study programs to educate scientists and engineers in these methods;
- directs NNI agencies to support the activities of standards setting bodies involved in the development of standards for nanotechnology, including authorizing agency reimbursement of travel costs of scientists and engineers participating in these activities; and
- requires activities supported under the NNI's Education and Societal Dimensions program component area to include environmental, health, and safety education in its informal, pre-college, and undergraduate nanotechnology education efforts.

H.R. 820—Nanotechnology Advancement and New Opportunities Act

H.R. 820, the Nanotechnology Advancement and New Opportunities Act, was introduced on February 3, 2009. Among its provisions, the bill would require the NNCO to produce an annual research strategy that establishes priorities for the development and responsible stewardship of nanotechnology, as well as providing recommendations regarding the funding required to implement the strategy.

On February 3, 2009, H.R. 820 was referred to the House Science and Technology Committee; the House Ways and Means Committee; the House Energy and Commerce Committee; and the House Homeland Security Committee.

APPENDIX A. SELECTED NANOTECHNOLOGY EHS ACTIVITIES OF FEDERAL REGULATORY AGENCIES

Several federal regulatory agencies have begun to grapple with the EHS issues raised by nanotechnology in their spheres of responsibility. Some critics argue that there is a potential conflict of interest among some regulatory agencies that are, on the one hand, conducting and promoting nanotechnology research and that are, on the other hand, responsible for regulating nanotechnology applications. The following section provides an overview of selected EHSrelated nanotechnology activities of federal regulatory agencies.

Environmental Protection Agency

The Environmental Protection Agency (EPA) co-chairs the NEHI working group of the NSET, along with the National Institute for Occupational Safety and Health (NIOSH), a research institute within the Department of Health and Human Services. EPA, which has both a research function and a regulatory function, has asserted a need for more information to assess the potential EHS impacts of most engineered nanoscale materials. According to EPA, this information is needed

> ... to establish a sound scientific basis for assessing and managing unreasonable risks that may result from the introduction of nanoscale materials into the environment.[61]

EPA is supporting research on the toxicology, fate, transport, transformation, bioavailability, and exposure of humans and other species to nanomaterials to obtain information for use in risk assessment, a central aspect of EPA's mission.[62]

EPA reports it is working collaboratively with stakeholders both domestically and internationally to address industrial chemical nanoscale materials. (International EHS collaboration is discussed in Appendix B.) One example of EPA's domestic work is its effort to establish a Nanoscale Materials Stewardship Program (NMSP). The purpose of the NMSP is to engage industry in a process that will foster effective federal government decision-making through the sharing of otherwise proprietary information about the characteristics, development, and manufacture of nanoscale materials. As envisioned by EPA, the program is designed primarily to engage manufacturers of nanoscale materials that would be considered existing chemical substances under the Toxic Substances Control Act (TSCA), but also encourages the participation of individuals and organizations working at a variety of stages of product development. EPA says that NMSP is intended to help provide a firmer scientific foundation for regulatory decisions by encouraging the development of key scientific information and appropriate risk management practices for nanoscale chemical substances.

According to EPA, the data acquired through NMSP will be used to gain an understanding of which nanoscale materials are produced, in what quantities, how they are used, and the data that are available for such materials. EPA maintains that its scientists will use data collected through this program, where appropriate, to aid in determining how and whether certain nanoscale materials or categories of nanoscale materials may present risks to human health and the environment. EPA states that NMSP is also intended to assist in the identification and adoption of risk management practices in the development and commercialization of nanoscale materials, to encourage the development of test data needed to provide a firmer scientific foundation for future work and regulatory/policy decisions, and to promote responsible development.[63]

EPA solicited comments on the NMSP from stakeholders in a July 2007 Federal Register Notice.[64] The business community has been supportive of the use of voluntary programs to address EHS risks of nanotechnology. The NanoBusiness Alliance states in its EHS research policy statement that "EPA and NIOSH should receive adequate funding to develop and implement their voluntary programs."[65] Other organizations have expressed frustration with the speed at which EPA is moving to implement the NMSP. At an EPA public meeting held in August 2007, Richard Denison, senior scientist for the Environmental Defense Fund, testified that

> As a government response to addressing the possible downsides of the nanotechnology revolution, [the NMSP is] simply 'too little, too late.'[66]

The Project on Emerging Nanotechnologies' J. Clarence Davies testified at the same meeting that while NMSP is

> ... potentially a useful initiative ... The delay in starting the NMSP is discouraging. It gives a signal that there really is no urgency, that the agency

> is in no hurry to start the voluntary program, much less institute an adequate regulatory system.[67]

Some observers say that past experience with other voluntary environmental programs shows that such efforts can produce benefits for both industry and government. For industry, voluntary programs may provide an opportunity to provide input into the regulatory process, to delay costly and constraining mandatory regulations, and to improve corporate goodwill. For government, voluntary programs may increase access to real-world data and information, may reduce the cost of data creation and/or collection, provide insights into new problems and about emerging industries, and provide a mechanism to control pollutants that are currently unregulated and for which jurisdiction may be hard to obtain.[68] Others maintain that voluntary programs can be counterproductive if they delay implementation of an adequate oversight system.

Multiple statutes govern EPA's authority to regulate nanotechnology materials and devices, including the Clean Air Act (CAA, 42 U.S.C. 7401 et seq); Clean Water Act (CWA, codified generally as 33 U.S.C. §§1251-1387); Federal Insecticide, Fungicide, and Rodenticide Act (FIFRA, 7 U.S.C.136-136y); and Toxic Substances Control Act (15 U.S.C. 2601 et seq.).[69] Important issues have been raised about the application of EPA's authorities to regulate nanotechnology.

Key issues revolve around TSCA, which authorizes regulation of chemical commerce.[70] Under the provisions of TSCA, producers of a "new" material must provide EPA with a premanufacture notification (PMN). EPA then has 90 days to approve manufacture, to require information from manufacturers, or to restrict chemical use. Other TSCA provisions permit EPA regulation of existing chemicals already in commerce, but these rely on EPA fact-finding and rulemaking before EPA can require testing or restrict uses. Several NGOs have urged EPA to consider all nanoscale materials "new" regardless of whether the material is on the EPA inventory list in its bulk form.[71] However, some nanotechnology materials have the same chemical composition as materials that are already in commerce, raising the question of whether the nanotechnology materials are "new" and thus subject to PMN requirements. With respect to this issue, EPA stated that

> EPA is considering how best to evaluate and, where appropriate, manage the risks associated with engineered nanoscale materials (NMs).... Nanoscale materials are "chemical substances" as defined under TSCA and are subject

to the law unless otherwise excluded. Thus premanufacture notifications (PMNs) are required under TSCA prior to manufacturing a NM "new" chemical substance. To assist potential submitters, EPA is developing a general approach to the TSCA inventory status of nanoscale substances in making the distinction between "new" and "existing" chemicals that are nanoscale materials. EPA is also developing an umbrella approach for evaluating both new and existing chemicals in NMs.[72]

When nanomaterials are intended to control pests, including microbes, FIFRA may offer EPA more authority to regulate nanotechnology than TSCA, according to Lynn Bergeson, chair of the American Bar Association's Section on Environment, Energy, and Resources:

> Under TSCA, once a substance is on the approved inventory list, any use is legitimate, but FIFRA is use-specific. The EPA always has the authority to assess the risk of pesticides, regardless of the use.[73]

Applicability of FIFRA to nanotechnology products was one aspect of a November 2006 EPA ruling that a device that "incorporates a substance intended to prevent, destroy or mitigate pests" is considered a pesticide and is required to be registered under FIFRA. While the ruling is not unique to nanomaterials, it came in the context of advertising claims for a washing machine containing nanoscale silver ions that kill microbes. EPA's ruling made this appliance the first nanotechnology product to be regulated under FIFRA. However, claims for the pesticidal effectiveness of the washing machine have been removed from advertisements, possibly limiting EPA's ability to regulate the device as a pesticide under FIFRA.

Food and Drug Administration

A variety of current and future products that incorporate nanotechnology fall, or may fall, under the regulatory auspices of the Food and Drug Administration (FDA), including cosmetics, medical devices, foods, drugs, biological products, and combination products.[74] FDA anticipates that many of the nanotechnology products that the agency is likely to regulate will be combination products, such as drug-device, drug-biological, or device-biological products. According to FDA, it regulates products based on their statutory classification rather than the technology they employ, thus the agency may not provide regulatory consideration to a nanotechnology product until well after its initial development.[75] Also, some critics maintain that FDA's limited regulatory authority over certain categories of products may limit its authority to regulate nanotechnology products.

With respect to the need for unique tests or requirements for regulating nanotechnology products, FDA states that its existing requirements may be adequate for most nanotechnology products it expects to regulate. FDA has asserted that nanotechnology products are in the same size-range as the cells and molecules its reviewers and scientists deal with every day. The agency says that every degradable medical device and injectable pharmaceutical generates particulates that pass through the nanoscale size range during the processes of their absorption and elimination by the body. According to FDA, it has no knowledge of reports of adverse reactions related to the "nano" size of resorbable drug or medical device products. New tests or other requirements may be needed, according to FDA, if new risks are identified arising from new materials or manufacturing techniques. FDA has established a Nanotechnology Interest Group (NTIG) comprised of representatives from each of its centers to facilitate the regulation of nanotechnology products.[76] Others, in particular consumer groups, counter that FDA's resources are insufficient to adequately address the safety of emerging technologies in general, and that the agency's regulatory approach, particularly for cosmetics, dietary supplements, and other products for which pre-market review is not required, would not detect any problems until such products had been in use.[77]

FDA does not provide grants for nanotechnology research but does conduct research in several of its centers to understand the characteristics of nanomaterials and nanotechnology processes. FDA is also collaborating with NIEHS on studies, as part of the interagency National Toxicology Program (NTP), examining the skin absorption and phototoxicity of nano-sized titanium dioxide and zinc oxide preparations used in sunscreens.

FDA says that there currently is no international regulation of nanoproducts or the underlying nanotechnology. FDA participates in multinational organizations where cooperative work on nanotechnology has been proposed, including the Organization for Economic Cooperation and Development (OECD), ASTM International, and the International Organization for Standardization (ISO). FDA plans to work with its foreign regulatory counterparts to share perspectives and information on regulation of nanotechnology.[78]

National Institute of Environmental Health Sciences/National Toxicology Program

While not a regulatory agency, NIEHS is conducting nanotechnology EHS research that will support the missions of regulatory agencies. In particular, NIEHS serves as home to the interagency National Toxicology Program. The NTP's mission is to coordinate toxicological testing programs, develop and validate improved testing methods, develop approaches and generate data to strengthen scientific knowledge about potentially hazardous substances, and communicate with stakeholders.[79] In 2006, the NTP established the Nanotechnology Safety Initiative (NSI), a broad-based research program to address potential human health hazards associated with the manufacture and use of nanoscale materials. The goal of this research program is to evaluate the toxicological properties of major nanoscale materials that represent a cross-section of composition, size, surface coatings, and physical and chemical properties, and to use these as model systems to investigate fundamental questions concerning whether nanoscale materials can interact with biological systems and how they might do so.[80]

According to NTP, the NSI is focused on three areas of research with respect to specific types or groups of nanoscale materials:

- non-medical, commercially relevant and available nanoscale materials to which humans are intentionally being exposed, such as cosmetics and sunscreens;
- nanoscale materials representing specific classes (e.g., fullerenes and metal oxides) so that information can be extrapolated to other members of those classes; and
- subsets of nanomaterials to test specific hypotheses about a key characteristic (such as size, composition, shape, or surface chemistry) that might be related to biological activity.

Current NSI research activities are focused on metal oxides, fluorescent crystalline semiconductors (also known as quantum dots), fullerenes, and carbon nanotubes.

NTP has also established a Nanotechnology Working Group (NWG) to serve as a technical advisory body to provide a structured and formal mechanism for bringing stakeholders together to learn about NTP nanotechnology research related to public health, address issues related to that research, and promote dissemination of those discussions to other federal agencies, nanotechnology stakeholders, and the public. Another function of the NWG is to provide a mechanism for the public and interested parties to provide advice to the NTP Board of Scientific Counselors.

Occupational Safety and Health Administration/National Institute for Occupational Safety and Health

The mission of the Occupational Safety and Health Administration (OSHA), an agency of the Department of Labor, is to ensure the safety and health of America's workers by setting and enforcing standards; providing training, outreach, and education; establishing partnerships; and encouraging continual improvement in workplace safety and health. OSHA has not yet taken any regulatory actions with respect to nanotechnology.

The National Institute for Occupational Safety and Health (NIOSH), a part of the Department of Health and Human Services, is the lead federal agency conducting research and providing guidance on the occupational safety and health implications and applications of nanotechnology. NIOSH co-chairs the NSET's NEHI working group. NIOSH is not a regulatory agency, but its work directly supports OSHA and other regulatory agencies.

NIOSH states that its nanotechnology efforts are building on its experience in defining the characteristics, properties, and effects of ultrafine particles—such as welding fumes and diesel particulates—as well as its experience in conducting advanced health effects laboratory studies and in fostering industrial hygiene policies and practices. NIOSH has developed interim guidelines for working with nanomaterials. The agency asserts that these guidelines are consistent with the best scientific knowledge of nanoparticle toxicity and control. NIOSH also maintains a Nanoparticle Information Library with information on the health and associated properties of nanomaterials as an online resource for occupational health professionals, industrial users, worker groups, and researchers.[81]

NIOSH and OSHA are considering risk management approaches that do not rely on traditional exposure- and time-limits. These new approaches seek to maximize flexibility for innovation while ensuring the health and safety of workers.[82]

Consumer Product Safety Commission

The Consumer Product Safety Commission (CPSC) is charged with protecting the public from unreasonable risks of serious injury or death from certain types of consumer products.[83] CPSC asserts that potential safety and health risks of nanomaterials can be assessed under existing CPSC statutes, regulations and guidelines. Since the Consumer Product Safety Act (15 U.S.C. 2051 et seq.) and the Federal Hazardous Substances Act (15 U.S.C. 1261 et seq.) do not require pre-market registration or approval of products, CPSC does not evaluate a product's risk to the public until it has been distributed in commerce.

In August 2005, CPSC commissioners approved a nanotechnology statement which notes that nanotechnology presents challenges that "may require unique exposure and risk assessment strategies." The CPSC statement identified regulatory challenges, including identification of the specific nanomaterial in a product; the need to characterize the materials to which a consumer is exposed during product use, including an assessment of the size distribution of the materials released; and the application of toxicological data of appropriate particle sizes to assess health risks. However, the CPSC takes the position that it is unable to make any general statements about potential consumer exposure to nanomaterials or the health effects that may result from exposure to nanomaterials during consumer use and disposal due to the wide variation in potential health effects and the dearth of exposure and toxicity data for specific nanomaterials.[84]

Appendix B. Selected International Engagement Efforts of NNI Agencies

Federal agencies have engaged internationally (e.g., with agencies of other nations, international organizations, standards organizations) on a host of nanotechnology-related issues, with a focus on EHS-related efforts such as scientific research, standards, nomenclature and terminology. The following section provides an overview of some of these activities.

In June 2004, the U.S. government initiated and hosted the first International Dialogue on Responsible Research and Development of Nanotechnology in Alexandria, Virginia. The meeting was attended by representatives from 25 countries and the European Union. The following year the NSET Subcommittee established the Global Issues in Nanotechnology (GIN) working group. In addition to monitoring foreign nanotechnology programs and promoting U.S. commercial and trade interests in nanotechnology, GIN was chartered to broaden international collaboration on nanotechnology R&D, including research on safeguarding the environment and human health. GIN representatives participated in the second International Dialogue on Responsible Research and Development of Nanotechnology hosted by the European Community (EC) in Brussels in July 2005. These meetings focused on clarifying issues and concerns of scientists, engineers, and policymakers working in nanotechnology around the world.

GIN representatives have also participated in nanotechnology-related activities of the Organization for Economic Cooperation and Development (OECD). In June 2005, chemical experts from 30 OECD countries participated in the Joint Meeting of Chemicals Committee and Working Party on Chemicals, Pesticides, and Biotechnology. Participants agreed to launch an international effort to coordinate assessment procedures for chemicals manufactured with nanotechnologies, to work toward linking national databases on high production-volume chemicals, and to establish a harmonized template for reporting hazard data needed for the notification and registration of new and existing chemicals, biocides, and pesticides. In December 2005, EPA hosted and chaired a second meeting of this group in Washington, D.C., on the safety of manufactured nanomaterials.

In October 2005, the United States proposed the creation of a Working Party on Nanotechnology within the OECD's Committee for Scientific and Technological Policy. Established in March 2007, the objective of this working party is to promote international co-operation that facilitates research, development, and responsible commercialization of nanotechnology in member countries and in non-member economies.[85] EPA is also participating in the OECD's Working Party on Manufactured Nanomaterials which was established in September 2006 to facilitate international collaboration on EHS issues related to manufactured nanomaterials.[86]

Another focus of U.S. international cooperation efforts has been in the development of nanotechnology standards. In response to a request from the White House Office of Science and Technology Policy, the American National Standards Institute (ANSI) established the Nanotechnology Standards Panel (NSP) in June 2004 to facilitate and coordinate nanotechnology standards development in the United States, focusing its initial work on nomenclature and terminology.[87] Subsequently, the International Organization for Standardization (ISO) established the Nanotechnologies Technical Committee, a parallel organization to ANSI's NSP. E. Clayton Teague, director of the NNI NNCO, chairs the ANSI-accredited Technical Advisory Group (TAG) to the ISO and leads the U.S. delegation.

The United States was selected to lead the ISO Technical Committee's Working Group on Health, Safety, and Environmental Aspects of Nanotechnologies. The Working Group has forwarded the NIOSH document "Approaches to Safe Nanotechnology," incorporating additional input from five other countries, to the ISO Technical Committee on Nanotechnologies (ISO TC 229) for a full review. If approved by the ISO Technical Committee, the document (re-titled "Health and Safety Practices in Occupational Settings Relative to Nanotechnologies") will be issued as an international Publicly Available Specification, an informational document available to all countries. NSET reports that significant progress on nanotechnology terminology and nomenclature has also been made by the TC 229 working group.

Federal government funding also contributed to the establishment of the International Council on Nanotechnology (ICON), a non-profit organization. ICON was established as an affiliate program of the NSF-funded, Rice University-based Center for Biological and Environmental Nanotechnology (CBEN), and has received funding from the National Science Foundation, the National Institutes of Health, and other private and non-profit organizations. ICON characterizes its purpose as seeking to catalyze global activities that lead to sound and responsible nanotechnology risk assessment, management, and communications. ICON has held EHS workshops, produced EHS reports, developed an online database of scientific findings related to the benefits and risks of nanotechnology, and designed and executed a survey of corporate nanotechnology EHS practices. In March 2007, ICON and CBEN jointly launched *The Virtual Journal of Nanotechnology Environment, Health, and Safety*, which contains citations and links to articles on the EHS impacts of nanotechnology.

End Notes

[1] Congress defined nanotechnology in the 21st Century Nanotechnology Research and Development Act (P.L. 108-153) as, "the science and technology that will enable one to understand, measure, manipulate, and manufacture at the atomic, molecular, and supramolecular levels, aimed at creating materials, devices, and systems with fundamentally new molecular organization, properties, and functions." ASTM International, one of the largest voluntary standards development organizations, has defined nanotechnology as, "A term referring to a wide range of technologies that measure, manipulate, or incorporate materials and/or features with at least one dimension between approximately 1 and 100

nanometers. Such applications exploit those properties, distinct from bulk or molecular systems, of nanoscale components." One nanometer is about the width of 10 hydrogen atoms placed side-by-side, or approximately 1/100,0000 of the thickness of a sheet of paper.

[2] Nanotechnology EHS applications refers to the beneficial use of nanotechnology to improve health, safety and the environment; EHS implications refers to known and potential adverse effects of nanoscale materials on health, safety and the environment.

[3] Project on Emerging Nanotechnologies. Figure as of April 2008.

[4] *Profiting From International Nanotechnology*, Lux Research, December 2006.

[5] "Genetically Modified Crops and Foods," Friends of the Earth, January 2003. http:// www.foe.co.uk/resource/ briefings/gm_crops_food.pdf

[6] "Fact Sheet for Nanotechnology under the Toxic Substances Control Act," Environmental Protection Agency. http://www.epa.gov/oppt/nano/nano-facts.htm

[7] "Cancer Nanotech Plan Gets Nod of Approval," Science, Vol. 305, July 23, 2004. http: //www.sciencemag.org/ content/vol305/issue5683/s-scope.dtl#305/5683/461c

[8] *A Matter of Size: Triennial Review of the National Nanotechnology Initiative,* National Research Council, 2006. p. 148.

[9] *Nanosilver: A Threat to Soil, Water and Human Health*? Friends of the Earth, March 2007. http://www.foe.org/pdf/ FoE_Australia_Nanosilver_report.pdf

[10] "Blood-Brain Barrier Breached by New Therapeutic Strategy," press release, National Institutes of Health, June 2007. http://www3.niaid.nih. gov/news/newsreleases/ 2007/bloodbrainbarrier.htm

[11] "Nanotechnology Risks: How Buckyballs Hurt Cells," Science Daily, May 27, 2008. http://www.sciencedaily.com/ releases/2008/05/ 080527091910.htm

[12] *National Nanotechnology Initiative: Research and Development Supporting the Next Industrial Revolution, Supplement to the President's FY2004 Budget*, Nanoscale Science, Engineering and Technology Subcommittee, Committee on Technology, National Science and Technology Council, The White House, October 2003. *http://www*nano.gov/nni04budget _supplement.pdf

[13] Magrez, A., Kasas, S., Salicio, V., Pasquier, N., Seo, J.W., Celio, M., Catsicas, S., Schwaller, B., and Forro, L. "Cellular Toxicity of Carbon-Based Nanomaterials," *Nano Letters*, 6(6):1121-1125, American Chemical Society, May 2006. http://pubs.acs.org/cgi-bin/abstract.cgi/nalefd/2006/6/i06/abs/nl060162e.htmll

[14] *Nanotechnology: The Future is Coming Sooner Thank You Think*, Joint Economic Committee, U.S. Congress, March 2007. p. 13. *http://www.house.gov/jec/publications/* 110/ nanotechnology_03-22-07.pdf

[15] The 21st Century Nanotechnology Research and Development Act (P.L. 108-153) requires a triennial assessment of the National Nanotechnology Program (in practice, of the NNI) by the NRC and a biennial assessment by PCAST, serving in its capacity as the National Nanotechnology Advisory Panel (NNAP). The act requires each assessment to include a review of the NNI's EHS activities. Three such assessments have been conducted, one by the NRC (*A Matter of Size: Triennial Review of the National Nanotechnology Initiative*, 2006) and two by PCAST (*The National Nanotechnology Initiative at Five Years: Assessments and Recommendations of the National Nanotechnology Advisory Panel*, May 2005; *The National Nanotechnology Initiative: Second Assessment and Recommendations of the National Nanotechnology Advisory Panel*, April 2008).

[16] *A Matter of Size: Triennial Review of the National Nanotechnology Initiative,* National Research Council, 2006. p. 90.

[17] *The National Nanotechnology Initiative: Second Assessment and Recommendations of the National Nanotechnology Advisory Panel*, President's Council of Advisors on Science and Technology, April 2008. p. 7.

[18] Bond, Phillip J., Under Secretary for Technology, U.S. Department of Commerce. "Preparing the Path for Nanotechnology: Addressing Legitimate Societal and Ethical Issues," keynote

address, Nanoscale Science, Engineering, and Technology Subcommittee Workshop on Societal Implications of Nanoscience and Nanotechnology, December 3, 2003.

[19] "Survey of U.S. Nanotechnology Executives," Small Times Magazine and the Center for Economic and Civic Opinion at the University of Massachusetts-Lowell, Fall 2006.

[20] Rejeski, David, Director, Project on Emerging Nanotechnologies. "Nanotech Safety 101 or How to Avoid the Next Little Accident," paper, Workshop on Disaster Prevention, Harvard University, April 27, 2006. http://www.nanotechproject.org/ file_download/files/ nanotechsafety101paper.pdf

[21] NEHI is a working group of the Nanoscale Science, Engineering, and Technology (NSET) Subcommittee of the White House National Science and Technology Council (NSTC). The NSET Subcommittee is the coordinating body for the NNI. For additional information about the structure of the NNI, see CRS Report RL34401, *The National Nanotechnology Initiative: Overview, Reauthorization, and Appropriations Issues*.

[22] According to the NNCO, EHS research funding data included in Tables 1 and 2 of this report are for implications research only. The NNCO also states that the figures reported in **Table 1** may understate the NNI's EHS implications research by excluding funding for instrument research, metrology, and standards that support EHS implications research but are reported separately. (Source: Private communication between the NNCO and CRS.)

[23] *The National Nanotechnology Initiative: Second Assessment and Recommendations of the National Nanotechnology Advisory Panel*, President's Council of Advisors on Science and Technology, April 2008. p. 34.

[24] *Prioritization of Environmental, Health, and Safety Research Needs for Engineered Nanoscale Materials*, Nanoscale Science, Engineering, and Technology Subcommittee, Committee on Technology, National Science and Technology Council, The White House, August 2007.

[25] *Environmental, Health, and Safety Research Needs for Engineered Nanoscale Materials,* Nanoscale Science, Engineering, and Technology Subcommittee, Committee on Technology, National Science and Technology Council, The White House, September 2006. Cover letter.

[26] Teague, E. Clayton, director, National Nanotechnology Coordination Office. Testimony before the Subcommittee on Research and Science Education, Committee on Science and Technology, U.S. House of Representatives. Hearing on "Research on Environmental and Safety Impacts of Nanotechnology: Current Status of Planning and Implementation under the National Nanotechnology Initiative." 110th Cong., 1st Sess., October 31, 2007.

[27] "Updated Administration Research and Development Budget Priorities," memorandum, Office of Management and Budget and Office of Science and Technology Policy, The White House, August 12, 2004. http://www.whitehouse.gov/ omb/memoranda/fy04/m04-23.pdf

[28] *A Matter of Size: Triennial Review of the National Nanotechnology Initiative,* National Research Council, 2006. p. 92.

[29] *The National Nanotechnology Initiative: Second Assessment and Recommendations of the National Nanotechnology Advisory Panel*, President's Council of Advisors on Science and Technology, April 2008. pp. 7, 27.

[30] Murdock, Sean, executive director, NanoBusiness Alliance. Testimony before the Committee on Science and Technology, U.S. House of Representatives. Hearing on "The National Nanotechnology Initiative Amendments Act of 2008." 110th Cong., 2nd Sess., April 16, 2008.

[31] Denison, Richard A. "A Proposal to Increase Federal Funding of Nanotechnology Risk Research to at least $100 Million Annually," Environmental Defense, April 2005. http://www.edf.org/documents/4442_100milquestionl.pdf

[32] *The National Nanotechnology Initiative: Second Assessment and Recommendations of the National Nanotechnology Advisory Panel*, President's Council of Advisors on Science and Technology, April 2008. pp. 7, 27.

[33] Kvamme, Floyd, co-chair, President's Council of Advisors on Science and Technology. Testimony before the Subcommittee on Research and Science Education, Committee on

Science and Technology, U.S. House of Representatives. Hearing on "Research on Environmental and Safety Impacts of Nanotechnology: Current Status of Planning and Implementation under the National Nanotechnology Initiative," 110th Cong., 1st Sess., October 31, 2007.

34 Maynard, Andrew, chief science advisor, Project on Emerging Nanotechnologies, a joint venture of the congressionally-chartered Woodrow Wilson Center for International Scholars and the Pew Charitable Trusts. Testimony before the Subcommittee on Research and Science Education, Committee on Science and Technology, U.S. House of Representatives. Hearing on "Research on Environmental and Safety Impacts of Nanotechnology: Current Status of Planning and Implementation under the National Nanotechnology Initiative," 110th Cong., 1st Sess., October 31, 2007.

35 The precautionary principle has been used in other countries on some issues and is the official policy in the European Union. For international agreements a precautionary approach is sometimes embraced. For example, the Biosafety Protocol to the 1992 Convention on Biological Diversity incorporates provisions applying the precautionary principle to the safe handling, transfer, and trade of genetically modified organisms. For further information, see CRS Report RL30594, *Biosafety Protocol for Genetically Modified Organisms: Overview*, by Alejandro E. Segarra and Susan R. Fletcher.

36 "NGOs urge precautionary principle in use of nanomaterials," EurActiv.com, June 14, 2007, http://www.euractiv.com/en/environment/ngos-urge-precautionary-principle-use-nanomaterials/article-164619; Sass, Jennifer, "Nanotechnology and the Precautionary Principle," presentation, Natural Resources Defense Council, 2006. http://docs.nrdc.org/health/hea_06121401A.pdf

37 *Nanomaterials, Sunscreens, and Cosmetics: Small Ingredients, Big Risks*, Friends of the Earth, May 2006. http://www.foe.org/camps/comm/nanotech/nanocosmetics.pdf

38 "No Small Matter," Communique, ETC Group, May/June 2002. *http://www.etc* group. org/upload/publication/ pdf_file/192 The ETC group is a non-governmental organization focused on the global societal impacts of emerging technologies.

39 "No Small Matter II: The Case for a Global Moratorium," Occasional Paper Series, ETC Group, April 2003. http://www.etcgroup.org/upload/publication/pdf_file/165

40 *Environmental, Health, and Safety Research Needs for Engineered Nanoscale Materials,* Nanoscale Science, Engineering, and Technology Subcommittee, Committee on Technology, National Science and Technology Council, The White House, September 2006. p. vii.

41 Bond, Phillip J., Under Secretary for Technology, U.S. Department of Commerce. "Nanotechnology: Economic Opportunities, Societal and Ethical Challenges," keynote address, NanoCommerce 2003, December 9, 2003.

42 *A Matter of Size: Triennial Review of the National Nanotechnology Initiative,* National Research Council, 2006. p. 92.

43 *The National Nanotechnology Initiative at Five Years: Assessments and Recommendations of the National Nanotechnology Advisory Panel*, President's Council of Advisors on Science and Technology, May 2005. p. 35.

44 *The National Nanotechnology Initiative: Second Assessment and Recommendations of the National Nanotechnology Advisory Panel*, President's Council of Advisors on Science and Technology, April 2008. p. 2.

45 Rejeski, David, director, Project on Emerging Nanotechnologies. Public comments on the Nanoscale Science, Engineering, and Technology Subcommittee's report, *Prioritization of Environmental, Health, and Safety Research Needs for Engineered Nanoscale Materials: An Interim Document for Public Comment*, September 12, 2007. *http://www*.nanotech project.org/process/files/5891/nehi_comments_070912_final.pdf

46 Hearing on "Research on Environmental and Safety Impacts of Nanotechnology: Current Status of Planning and Implementation under the National Nanotechnology Initiative." 110th Cong., 1st Sess., October 31, 2007.

[47] An electronic copy of this letter, dated February 22, 2007, was provided to the Congressional Research Service (CRS) by the American Chemistry Council.

[48] Incorporated as division F of the Consolidated Appropriations Act, 2008 (P.L. 110-161).

[49] S.Rept. 110-91, p. 54.

[50] Maynard, Andrew, chief science advisor, Project on Emerging Nanotechnologies. "Public Meeting on Research Needs and Priorities Related to Environmental, Health, and Safety Aspects of Engineered Nanoscale Materials," comments, January 4, 2007. http://www.nano.gov/html/meetings/ehs/uploads/ 20070103_1505_Nanotechnology_Maynard_NNCO_Comments.pdf

[51] Teague, E. Clayton, director, National Nanotechnology Coordination Office. Testimony before the Subcommittee on Research and Science Education, Committee on Science and Technology, U.S. House of Representatives. Hearing on "Research on Environmental and Safety Impacts of Nanotechnology: Current Status of Planning and Implementation under the National Nanotechnology Initiative." 110th Cong., 1st Sess., October 31, 2007.

[52] E-mail communication, November 21, 2007.

[53] "Nanotechnology Leaders to Converge in Washington, D.C., This Week for NanoBusiness Alliance Public Policy Tour," article, nanotechwire.com, February 16, 2006. http://nanotechwire.com/news.asp?nid=2929

[54] Davies, J. Clarence. *Managing the Effects of Nanotechnology*, Project on Emerging Nanotechnologies, January 2006. p. 3. http://www.nanotechproject.org/ process/assets

[55] "U.S. Risks Losing Nano Lead," article, physorg.com, July 6, 2005. http:// www.physorg.com/news4963.html

[56] "International Coalition Calls for Oversight of Nanotechnology," press release, Friends of the Earth, July 31, 2007. http://action.foe.org/dia/organizations ORG/foe/press Release. jsp?press_release_KEY=248

[57] Ibid.

[58] *A Matter of Size: Triennial Review of the National Nanotechnology Initiative,* National Research Council, 2006. p.11.

[59] *The National Nanotechnology Initiative: Second Assessment and Recommendations of the National Nanotechnology Advisory Panel*, President's Council of Advisors on Science and Technology, April 2008. p. 33.

[60] *Regional, State, and Local Initiatives in Nanotechnology*, Nanoscale Science, Engineering, and Technology Subcommittee, Committee on Technology, National Science and Technology Council, The White House, 2005. p. 33.

[61] "Fact Sheet for Nanotechnology under the Toxic Substances Control Act," Environmental Protection Agency. http://www.epa.gov/oppt/nano/nano-facts.htm

[62] "Exploratory Research: Nanotechnology Research Grants Investigating Fate, Transport, Transformation, and Exposure of Engineered Nanomaterials: A Joint Research Solicitation - EPA, NSF, & DOE," Environmental Protection Agency. http://es.epa.gov/ncer/rfa/ 2007/2007 _star_nanotech.html

[63] EPA notes that the National Research Council described "responsible development" in its first triennial review of the NNI as "the balancing of efforts to maximize the technology's positive contributions and minimize its negative consequences. Thus, responsible development involves an examination both of applications and of potential implications. It implies a commitment to develop and use technology to help meet the most pressing human and societal needs, while making every reasonable effort to anticipate and mitigate adverse implications or unintended consequences."

[64] "Nanoscale Program Approach for Comment," Environmental Protection Agency. http://www.epa.gov/oppt/nano/ nmspfr.htm

[65] "Nanotech Environmental, Health and Safety: Progress and Priorities," NanoBusiness Alliance. http://www.nanobusiness.org/ehspolicy.php

[66] "EPA's Actions on Health Risks of Nanomaterials Called 'Too Little, Too Late,'" press release, Environmental Defense Fund, August 2, 2007.

[67] Davies, J. Clarence, testimony, EPA Public Meeting on Nanoscale Materials Stewardship Program, August 2, 2007. http://www.nanotechproject.org/file_download/212

[68] Berger, Michael. "Implementing successful voluntary nanotechnology environmental programs appears to be a challenge," Nanowerk LLC, November 29, 2007. http://www.nanowerk.com/spotlight/spotid=3476.php

[69] For additional information, see CRS Report RL30798, *Environmental Laws: Summaries of Major Statutes Administered by the Environmental Protection Agency (EPA)*, by Susan R. Fletcher et al.

[70] For more information about TSCA and nanotechnology, see CRS Report RL34118, *The Toxic Substances Control Act (TSCA): Implementation and New Challenges*, by Linda-Jo Schierow.

[71] "The EPA's Toxic Substances Control Act: What you must know," *Small Times*, September/October 2007.

[72] "New Nanotechnology Products," Environmental Protection Agency. *http://www.epa.* gov/oppt/ar/20052006/ managing/new_nano.htm

[73] "EPA Regulates Nano Product, Not Nano Industry," *Small Times*, January 2007.

[74] For additional information, see CRS Report RL34334, *The Food and Drug Administration: Budget and Statutory History, FY1980-FY2007*, coordinated by Judith A. Johnson.

[75] "FDA and Nanotechnology Products," Food and Drug Administration. http://www.fda.gov/nanotechnology/faqs.html

[76] Ibid.

[77] See, for example, Michael Taylor, *Regulating the Products of Nanotechnology: Does FDA Have the Tools It Needs?* The Project on Emerging Nanotechnologies, October 2006, at http://www.nanotechproject.org/news/archive/ is_fda_nanotech-ready. The FDA Science Board, Subcommittee on Science and Technology, designated nanotechnology as one of eight emerging technologies that are most challenging for FDA. See FDA Science Board, Subcommittee on Science and Technology, *FDA Science and Mission at Risk*, November 2007, p. 4, at http://www.fda.gov/ohrms/dockets/ac/07/briefing/2007-4329b_02_01_FDA Report on Science and Technology.pdf. The FDA Science Board is the advisory board to the FDA Commissioner.

[78] *FDA and Nanotechnology Products: Frequently Asked Questions*, Food and Drug Administration, U.S. Department of Health and Human Services. http://www.fda.gov/nanotechnology/faqs.html

[79] "Toxicology in the 21st Century: The Role of the National Toxicology Program," Update, National Toxicology Program, January 2004. http://ntp.niehs.nih.gov/ntp/htdocs/Liaison/2004JanLO_News.pdf

[80] *Toxicology in the 21st Century: The Role of the National Toxicology Program,* Department of Health and Human Services, February 2004. http://ntp.niehs.nih.gov/ntp/mainpages /NTPVision.pdf

[81] "Nanotechnology at NIOSH," National Institute for Occupational Safety and Health. http://www.cdc.gov/niosh/ topics/nanotech/

[82] *Environmental, Health, and Safety Research Needs for Engineered Nanoscale Materials,* Nanoscale Science, Engineering, and Technology Subcommittee, Committee on Technology, National Science and Technology Council, The White House, September 2006.

[83] CPSC's regulatory authorities are provided by the Consumer Product Safety Act; Federal Hazardous Substances Act of 1960, as amended by the Toy Safety Acts of 1969 and 1984 and the Child Protection Amendments of 1966; Poison Prevention Packaging Act of 1970; Flammable Fabrics Act of 1953; and Refrigerator Safety Act of 1956. For additional information, see CRS Report RS22821, *Consumer Product Safety Commission: Current Issues*, by Bruce K. Mulock.

[84] "CPSC Nanomaterial Statement," Consumer Product Safety Commission, August 2005. http://www.cpsc.gov/library/ cpscnanostatement.pdf

[85] "OECD Work on Nanotechnology," Organization for Economic Cooperation and Development. http://www.oecd.org/sti/nano

[86] "Fact Sheet for Nanotechnology Under the Toxic Substances Control Act," Environmental Protection Agency. http://www.epa.gov/oppt/nano/nano-facts.htm; "OECD Work on Nanotechnology," Organization for Economic Cooperation and Development. http://www.oecd.org/sti/nano

[87] "ANSI Establishes Nanotechnology Standards Panel," American National Standards Institute, August 5, 2004. http://www.ansi.org/news_publications/ news_story.aspx? menuid =7& articleid=735

In: Nanotechnology Considerations for the EPA... ISBN: 978-1-61761-817-8
Editors: Scott J. Castiglione

Chapter 3

NANOTECHNOLOGY: A REPORT OF THE U.S. FOOD AND DRUG ADMINISTRATION

Nanotechnology Task Force

EXECUTIVE SUMMARY

As other emerging technologies have in the past, nanotechnology poses questions regarding the adequacy and application of regulatory authorities. The then Acting Commissioner of the Food and Drug Administration (FDA) initiated the Nanotechnology Task Force (Task Force) in 2006 to help assess these questions with respect to FDA's regulatory authorities, in light of the current state of the science for nanotechnology. This report offers the Task Force's initial findings and recommendations to the Commissioner.

The report includes:

- A synopsis of the state of the science for biological interactions of nanoscale materials;
- Analysis and recommendations for science issues; and
- Analysis and recommendations for regulatory policy issues.

The report addresses scientific issues as distinct from regulatory policy issues in recognition of the important role of the science in developing regulatory policies in this area, rapid growth of the field of nanotechnology, and the evolving state of scientific knowledge relating to this field. Rapid developments in the field mean that attention to the emerging science is needed to enable the agency to predict and prepare for the types of products FDA may see in the near future.

A general finding of the report is that nanoscale materials present regulatory challenges similar to those posed by products using other emerging technologies. However, these challenges may be magnified both because nanotechnology can be used in, or to make, any FDA-regulated product, and because, at this scale, properties of a material relevant to the safety and (as applicable) effectiveness of FDA-regulated products might change repeatedly as size enters into or varies within the nanoscale range. In addition, the emerging and uncertain nature of the science and potential for rapid development of applications for FDA-regulated products highlights the need for timely development of a transparent, consistent, and predictable regulatory pathway.

The Task Force's initial recommendations relating to scientific issues focus on improving scientific knowledge of nanotechnology to help ensure the agency's regulatory effectiveness, particularly with regard to products not subject to premarket authorization requirements. The report also addresses the need to evaluate whether the tools available to describe and evaluate nanoscale materials are sufficient, and the development of additional tools where necessary.

The Task Force also assessed the agency's regulatory authorities to meet any unique challenges that may be presented by FDA-regulated products containing nanoscale materials. This assessment focused on such broad questions as whether FDA can identify products containing nanoscale materials, the scope of FDA's authorities to evaluate the safety and effectiveness of such products, whether FDA should require or permit products to be labeled as containing nanoscale materials, and whether the use of nanoscale materials in FDA-regulated products raises any issues under the National Environmental Policy Act.

The Task Force concluded that the agency's authorities are generally comprehensive for products subject to premarket authorization requirements, such as drugs, biological products, devices, and food and color additives, and that these authorities give FDA the ability to obtain detailed scientific information needed to review the safety and, as appropriate, effectiveness of products. For products not subject to premarket authorization requirements, such as dietary supplements, cosmetics, and food ingredients that are generally recognized as safe (GRAS), manufacturers are generally not required to submit data to FDA prior to marketing, and the agency's oversight capacity is less comprehensive.

The Task Force has made various recommendations to address regulatory challenges that may be presented by products that use nanotechnology, especially regarding products not subject to premarket authorization requirements, taking into account the evolving state of the science in this area. A number of recommendations deal with requesting data and other information about effects of nanoscale materials on safety and, as appropriate, effectiveness of products. Other recommendations suggest that FDA provide guidance to manufacturers about when the use of nanoscale ingredients may require submission of additional data, change the product's regulatory status or pathway, or merit taking additional or special steps to address potential safety or product quality issues. The Task Force also recommends seeking public input on the adequacy of FDA's policies and procedures for products that combine drugs, biological products, and/or devices containing nanoscale materials to serve multiple uses, such as both a diagnostic and a therapeutic intended use. The Task Force also recommends encouraging manufacturers to communicate with the agency early in the development process for products using nanoscale materials, particularly with regard to such highly integrated combination products.

The guidances the Task Force is recommending would give affected manufacturers and other interested parties timely information about FDA's expectations, so as to foster predictability in the agency's regulatory processes, thereby enabling innovation and enhancing transparency, while protecting the public health.

INTRODUCTION

Nanoscale materials often have chemical, physical, or biological properties that are different from those of their larger counterparts. Such differences may include altered magnetic properties, altered electrical or optical activity, increased structural integrity, or altered chemical or biological activity. Because of these properties, nanoscale materials have great potential for use in a vast array of products. Of particular interest to the Food and Drug Administration (FDA, the agency), nanoscale materials may enable new developments in products to advance public health. Also because of some of their special properties, nanoscale materials may pose different safety issues than their larger or smaller (i.e., molecular) counterparts.

FDA is generally responsible for overseeing the safety and effectiveness of drugs and devices for humans and animals, and of biological products for humans. The agency is also generally responsible for overseeing the safety of foods (including food additives and dietary supplements), color additives, and cosmetics.[1] The agency conducts these oversight functions under a variety of laws and regulations, which establish the specific pre-market and/or post-market oversight mechanisms applicable to a particular class of products. Most of the laws and regulations under which FDA operates were written before the advent of nanotechnology. Most are general in nature by design, however, offering flexibility to accommodate products made with new technologies or containing new kinds of materials.

Research and development relating to nanotechnology applications promises the development of products having multiple, highly integrated functions. FDA will need to anticipate this shift in the nature of products received for review and authorization. For example, disease diagnosis, drug targeting, and non-invasive imaging elements are being combined in individual nanotechnology products.[2] A goal of this report is to assist in the development of a transparent, consistent, and predictable regulatory pathway for such products.

More broadly, this report is intended to outline ways in which FDA can both (1) enhance its knowledge of nanotechnology to support its oversight for products using such technology and (2) inform interested stakeholders of what information may need to be developed to support the marketing of FDA-regulated products that use nanoscale materials.

FDA is a member agency in the National Nanotechnology Initiative (NNI), a federal research and development program established to coordinate the multi-agency efforts in nanoscale science, engineering, and technology. The goals of the NNI are to: (1) maintain a world-class research and development program aimed at realizing the full potential of nanotechnology; (2) facilitate transfer of new technologies into products for economic growth, jobs, and other public benefit; (3) develop educational resources, a skilled workforce, and the supporting infrastructure and tools to advance nanotechnology; and, (4) support responsible development of nanotechnology. FDA centers are conducting nanoscale material research within programs that support their specific regulatory priorities. Participation in the NNI provides FDA and the other regulatory and research funding agencies the opportunity to define their research needs.

Task Force Mission

In August 2006, then Acting Commissioner of Food and Drugs, Andrew C. von Eschenbach, M.D., announced the formation of an internal FDA Nanotechnology Task Force. He charged the Task Force with determining regulatory approaches that would enable the continued development of innovative, safe, and effective FDA-regulated products that use nanoscale materials.[3] The Task Force was asked to identify and recommend ways to address any knowledge or policy gaps that exist to better enable the agency to evaluate safety aspects of FDA-regulated products that contain nanoscale materials. Specifically, the Task Force was directed to:

- Chair a public meeting to help FDA further its understanding of developments in nanoscale materials that pertain to FDA-regulated products, including new and emerging scientific issues such as those pertaining to biological interactions that may lead to either beneficial or adverse health effects;
- Assess the current state of scientific knowledge pertaining to nanoscale materials for purposes of carrying out FDA's mission;
- Evaluate the effectiveness of the agency's regulatory approaches and authorities to meet any unique challenge that may be presented by the use of nanoscale materials in FDA-regulated products;
- Explore opportunities to enable innovation using nanoscale materials to develop safe and effective drugs, biologics and devices, and to develop safe foods, feeds, and cosmetics;
- Continue to strengthen FDA's collaborative relationships with other federal agencies, including the agencies participating in the NNI such as the National Institutes of Health (NIH), the Environmental Protection Agency (EPA), and the United States Department of Agriculture (USDA), as well as with foreign government regulatory bodies, international organizations, healthcare professionals, industry, consumers, and other stakeholders, to gather information regarding nanoscale materials used or that could be used in FDA-regulated products;
- Consider appropriate vehicles for communicating with the public about the use of nanoscale materials in FDA-regulated products; and
- Submit its initial findings and recommendations to the Commissioner within nine months of the public meeting.

As requested by the Commissioner, the Task Force opened a public docket and held a public meeting on October 10, 2006. The objectives of the meeting and the docket were to learn about:

- New nanoscale material products under development in the areas of foods (including food additives and dietary supplements), color additives, animal feeds, cosmetics, drugs and biologics, and medical devices;
- New or emerging scientific issues that should be brought to FDA's attention, including issues related to the safety of nanoscale materials; and
- Any other issues concerning the use of nanoscale materials in FDA-regulated products regarding which regulated industry, academia and the interested public wished to inform FDA.

Following the public meeting, the Task Force:

- Reviewed both the oral and written comments it received;
- Assessed the current state of nanotechnology-related science; and
- Evaluated the scope of the agency's current authorities to meet any unique challenges presented by the use of nanoscale materials in FDA-regulated products.

Definitions for Nanotechnology and Related Terms

Nanotechnology allows scientists to work on the scale of molecules to create, explore, and manipulate the biological and material worlds measured in nanometers, one-billionth of a meter. By way of comparison, a sheet of paper is about 100,000 nanometers thick; a human hair is about 80,000 nanometers wide.

FDA regulates a wide range of products, which may use a wide range of materials for varied purposes inside and outside the body. The Task Force has sought to apply a sufficiently comprehensive analytical approach to address the scientific and policy issues FDA must consider to protect and promote human and animal health in regulating products within its jurisdiction that use nanoscale materials. Accordingly, the Task Force has considered a broad array of available information on a wide range of products, materials, techniques, and technologies. The Task Force has not adopted a precise definition for "nanoscale materials," "nanotechnology," or related terms to define the scope of its work. The Task Force concluded that it would be most productive to take a broadly inclusive approach in identifying potentially relevant studies, data, and other information.

The Task Force believes FDA should continue to pursue regulatory approaches that take into account the potential importance of material size and the evolving state of the science. Moreover, while one definition for "nanotechnology," "nanoscale material," or a related term or concept may offer meaningful guidance in one context, that definition may be too narrow or broad to be of use in another. Accordingly, the Task Force does not recommend attempting to adopt formal, fixed definitions for such terms for regulatory purposes at this time. As FDA learns more about the interaction of nanoscale materials with biological systems and generalizable concepts that can inform the agency's judgment, it may be productive to develop formal, fixed definitions, appropriately tailored to the regulation of nanoscale materials in FDA-regulated products.[4]

Organization of this Report

This report provides the Task Force's initial findings and recommendations. The report is divided into three sections. The first section discusses scientific knowledge of the potential effects of nanoscale materials relevant to FDA's regulation of products. The second section assesses scientific issues relating to FDA's regulation of products using nanoscale materials. The third section assesses the agency's regulatory authorities as these authorities relate to FDA-regulated products using nanoscale materials. The report takes into account comments submitted to relevant public dockets and made at the public meeting, and then presents recommendations to the Commissioner for actions the agency can take in furtherance of its mission to protect and promote the public health.

State of the Science of Nanotechnology Relevant to FDA

The science and applications of nanotechnology are developing at a very rapid pace. In 1990, approximately 1,000 scientific publications on nanotechnology were published and approximately 200 patent applications (worldwide) were filed.[5] By 2002, the number of publications had risen to over 22,000 with over 1,900 patent applications. This exponential increase in scientific publications and patents is the result of increased discovery and investment in nanotechnology that will likely result in substantial and continual changes in products falling under the regulatory authority of the FDA.

Numerous reviews have been published focusing on the state of knowledge of behavior, interaction with biological systems (both for advantageous and toxicity outcomes), and potential environmental disposition of nanoscale materials. Government bodies have published some of these,[6] as have private entities focusing on the risk associated with nanoscale materials,[7] and still others can be found in the peer-reviewed public literature.[8] Often these reports stress the need for research on: characterization of materials in the nanoscale range; methods for identifying hazards; understanding biological response to nanoscale materials; and characterizing nanoscale material exposure and transport (in humans and the environment).

Identifying precisely what qualifies as a nanoscale material is difficult and currently a subject of substantial discussion in the scientific, regulatory, and standards communities. As a result, developing a comprehensive description of products that are currently produced with nanotechnology, or may be produced with this technology in the future, would be difficult at best, and likely infeasible. Instead this report considers examples based on what is currently known about use of this technology.

Perhaps of greatest relevance to products regulated by FDA is what the current study of structures in the nanoscale range is teaching about biological interactions.

In light of the current state of the science, an understanding of the interactions of nanoscale materials with biological systems is probably best gained through case-by-case analysis of specific types of nanoscale materials and of variations in their characteristics. However, recent reviews have developed initial approaches for more general description of the kinds of interactions that might be expected with biological systems.[9] Generalized approaches to organizing information for risk assessment and risk management of nanoscale materials have also been developed that provide insight for how biological interactions of nanoscale materials might be understood.[10]

Generalizable Knowledge of Biological Interactions

Generalizable principles are being derived from studies of specific types of nanoscale materials. For instance, studies in some laboratories have determined that surface area per unit of volume or mass can be a better measure than mass for assessing relative toxicity across particle size and material variations.[11] This tells us, for example, that the suitability of mass-based dose measurements should be evaluated before drawing conclusions about potency of a drug effect or about toxic response. However, it is important to recognize that in some cases surface area has been shown to be a much less important determinant of biological interaction than surface modification (for example, binding different chemical groups to the surface of a particle) for those particular types of particles in which this has been examined systematically.[12] This indicates that one should pay particular attention to the composition and surface characteristics of nanoscale materials that may come in contact with biological systems.

In some cases, features of materials such as charge or the position of the surface modification have been found to affect toxicity.[13] For example, positively charged nanoscale lipid vesicles (nanovesicles) induced cerebral edema, but neutral nanovesicles and low concentrations of negatively charged nanovesicles did not.[14] Studies have also shown that modifying the surface of nanoscale materials with surfactants or biocompatible polymers (e.g., polyethylene glycol, PEG) reduced the toxicity *in vitro*[15] and altered the half-life and tissue deposition *in vivo*.[16] Such findings are relevant to drug delivery for example, for understanding the potential distribution of nanoscale materials in the body, and for evaluating toxicity and biocompatibility. However, these findings are material-specific and, we are not aware of methods or models that would extend these findings to provide a better understanding of broad classes of materials.

There is also well-developed literature on biological interactions of naturally occurring particles or particles released from industrial processes that include particles in the nanoscale range, such as combustion-related particulate matter, silica dust, and biological particles. These data sets may provide valuable information for understanding generalizable properties of nanoscale particles. However, because of the great significance of surface modification to the properties of nanoscale particles that has been shown in some cases, the value of these existing data that deal with widely variable nanoscale particle subtypes may be limited to understanding of basic issues such as biological filtering and dose metrics.

Predictability of Biological Interactions

At a first level of approximation, understanding simple particle movement into or from one compartment to another in the body involves consideration of various absorption and transport mechanisms that either passively keep particles from entering some areas of the body depending on how large they are, or actively move the particles. An understanding of these mechanisms can help predict what movement might occur for particular nanoscale particles. Accordingly, the significance of size to biological interaction may be relatively straightforward to predict and assess in many cases where changing particle size can be expected to affect absorption and transport mechanisms.

Furthermore, if a specific kind of surface reactivity is known for a material, then it might be expected that the reactivity per unit of mass would increase with decreasing particle size because specific surface area would increase. It could also, therefore, be predicted that important biological effects could arise when the surface area is substantially increased with the same mass of exposure. The argument has been made, for example, that generation of reactive oxygen species for some particles may increase as particle size decreases and surface area increases, leading to increased inflammatory response in lung tissue.[17] The particle size range or particle concentrations at which such an increase in reactivity could cause adverse effects would depend on adaptive responses of the affected biologic system, however, and, therefore, would be difficult to predict in the absence of test data.

In addition, biological interaction may be focused such that only a specific particle size range will have biological effects. For example, there may be an upper size limit for dermal absorption of any specific type of particle, though that limit may vary based on other factors as well (e.g., particle charge or chemical properties, conditions of dermal exposure). A compound that previously was excluded from exposure to internal tissues might then produce effects, beneficial or adverse, due to the exposure of internal tissues when particle size decreased below the threshold for exclusion. These effects could be based on somewhat simple physical interaction, such as the filtering capacity of phagocytic cells, transport effects of capillary structures, or adhesion to proteins or other molecules in biological fluids.[18] In some cases, these interactions may be predictable, for example, based on knowledge of the size of materials that can pass through capillary walls. In others, data could be developed that would allow better prediction of biological interaction, for example regarding size-dependent dermal absorption as is being developed by FDA in collaboration with the National Toxicology Program.[19]

Biological interactions influenced by the particular chemistry and physical configuration of the nanoscale material might also occur in ways that are unpredictable without specific test data for the material. For example, similar to how charge and functional group locations on a molecule will affect interactions with chemicals in the body, characteristics such as a particle's shape and the location of changes in its surface may affect the interactions of nanoscale materials with chemicals in the body.

The available information does not suggest that all materials with nanoscale dimensions will be hazardous. Furthermore, if all nanoscale materials are compared to all non-nanoscale materials, whether larger or smaller, it is not apparent that the nanoscale materials as a group would have more inherent hazard. However, consideration of the basic science of how materials interact with biological systems does indicate that a material's properties can change when size is increased or decreased into, or varied within, the nanoscale range.

Science Issues

Introduction

Virtually any product category regulated by FDA might currently or in the future involve uses of nanotechnology somewhere in the manufacturing process. A somewhat smaller set of products can be expected to retain nanoscale structures in the finished product, such as systemically bioavailable nanoscale drugs or nanoscale particles or nanoscale structures in solid objects that may release nanoscale materials through use. As discussed above, the biological interactions of regulated products using nanoscale materials are of particular importance to FDA. This section addresses scientific issues relevant to FDA. They relate to:

- Understanding of interactions of nanoscale materials with biological systems; and
- Adequacy of testing approaches for assessing safety, effectiveness, and quality of products containing nanoscale materials.

We will address both issues below, taking into account relevant comments submitted to the public docket for the FDA Nanotechnology Task Force's October 2006 public meeting and to the docket opened in response to a petition submitted by the International Center for Technology Assessment

(ICTA) and other parties entitled, Petition to FDA to Amend its Regulations for Products Composed of Engineered Nanoparticles Generally and Sunscreen Drug Products Composed of Engineered Nanoparticles Specifically (ICTA Petition).[20] In addition to other resources, the Task Force also considered the US Government-wide evaluation of Environmental, Health, and Safety Research Needs for Engineered Nanoscale Materials (NNI 2006c)[21] in developing this discussion and these recommendations.

Issue: Understanding Interactions of Nanoscale Materials with Biological Systems

Comments

Many comments noted that nanoscale materials can have a wide variety of properties that are markedly different from the same materials in larger scale forms. Some comments suggested that the definition of "nanomaterial" should be limited to those materials that have some unique, qualitatively different, properties derived from their physical scale. Some comments noted that although the nature and unique properties of many nanoscale materials are not well understood, some nanoscale materials have been observed to be toxic in certain assays and under some specified conditions, or, based on their behavior in biological systems, raise suspicions of potential toxicity. .

Some comments stated that nanoscale materials have a unique ability to interact with proteins and other essential biological functional elements. Some noted: that nanoscale materials can be more biologically active than non-nanoscale materials; that basic research is needed on such issues as interactions with subcellular structures and dose/concentration; and that such research should take an interdisciplinary approach, making use of experts in toxicology, materials science, medicine, molecular biology and bioinformatics. The comments pointed out that there are differences in dose-response curves depending on whether the curves are expressed by mass, number of particles, or surface area.

Some comments essentially stated that, because properties or safety of nanoscale materials cannot be assumed or inferred from larger-scale forms, nanoscale materials will need to be directly and adequately evaluated. Some comments highlighted that this lack of proven safety may be a particularly significant concern for ingested products.

Other comments noted that there is a great deal of knowledge about behavior of many types of nanoscale materials, that there are established methods for assessing their safety and effectiveness, and that the novelty of the scientific issues posed by nanoscale materials is exaggerated. One comment stated that the nanoscale materials contained in some newer sunscreens have been well tested by industry already and have had their safety established. Some comments noted that the use of nanoscale materials may enable development of more targeted drugs and biological products, which may be safer and more effective than otherwise possible, and development should not be impeded.

Some comments highlighted the need for FDA to have adequately trained and educated staff to review products that may contain nanoscale materials.

Analysis

Issues Relevant to all Regulated Products

There may be a fundamental difference in the kind of uncertainty associated with nanoscale materials compared to conventional chemicals, both with respect to knowledge about them and the way that testing is performed. For conventional chemicals, there is a relatively long history of exploring, and a correspondingly relatively robust understanding, of interactions of molecular classes (such as compounds with particular structures or functional groups) with biological systems. In some cases, screening test methods are used to define what additional tests may need to be performed to gain sufficient knowledge about safety and/or effectiveness. For example, there are screening tests available to help identify whether DNA damage is a possible outcome from exposure to a certain chemical. Other tests can tell how the chemical is distributed in the body and in what forms it is present in various tissues. FDA has an expectation relevant to molecular forms of materials used in products that FDA regulates that if the molecule does not cause DNA damage during *in vitro* testing, or if it is metabolized quickly and does not reach sensitive organs, or if it is not absorbed, then it is less likely to present certain kinds of health hazards. This expectation is based on long experience with, and consequent understanding of, basic biological interactions of molecular forms of chemicals and of how these interactions correlate with the results of current testing methods.

The testing methods for different product types also may need to be evaluated to determine whether or how they can be used in assessing the bioavailability of nanoscale materials in humans. This issue is not so much one of whether the tests are valid (as discussed under "Adequacy of testing approaches…" below). Rather, it speaks to the familiarity with the predictive value of the tests. There is, comparatively speaking, more familiarity with the predictive value of such tests for molecules than for nanoscale materials. As FDA and the scientific community develop familiarity with different nanoscale material types, it may be possible to predict, for example, that specific variations in characteristics of a type of nanoscale material may cause it to be reactive in the same way that it is known that addition of a chemical-specific functional group may cause a molecule to be reactive.

Several recent scientific reviews conclude that the state of knowledge for biological interactions of nanoscale materials is generally in need of improvement to enhance risk assessments and better support risk management decisions.[22] For example, the information available on biological interactions for nanoscale materials primarily applies to the specific materials tested, and the agency is not aware of models for how the information might be applied more broadly to other nanoscale materials. At this stage, it is important to monitor individual hazard studies of specific nanoscale materials and, where possible, seek to synthesize a common understanding of more general material properties from these studies.

There is a potential to develop and organize information using physiologically based pharmacokinetic (PbPk) models or quantitative structure activity relationship (QSAR) models, to enable formulation of generalized principles for the behavior of classes of nanoscale materials. Integration of information with respect to given material characteristics (for example, material type, size, charge, surface modification) could be of particular benefit in improving general understanding. Such models are useful in predicting biological interactions of molecules and may provide the same sort of value for nanoscale materials. At present, however, the agency is not aware of models such as these for nanoscale materials.

Further, even though there is a broad array of ongoing research across a wide variety of disciplines, the different disciplines and laboratories employ different methods, standards and ontologies. As mentioned previously, in 2002 alone there were 22,000 scientific publications relating to nanotechnology.[23] Managing this information is a formidable task, but data mining would likely yield a wealth of information relevant to the FDA's review of nanotechnology-based products. For these reasons, the development of a comprehensive database employing standardized ontologies, or other means of integration, could be very valuable.

Rendering appropriate regulatory decisions requires up-to-date training and information. FDA's ability to accomplish its mission depends in part on having staff with expertise in areas such as pharmacology, materials science, biology, physics, chemistry, medicine, and toxicology. New and emerging nanotechnology-based products highlight the possible need for new expertise for some review areas. For example, characterization methods for nanoscale materials (e.g., describing size, shape, surface topography) generally require use of different equipment than would be used for characterizing molecules. The data produced by this equipment are different than those generated by chemical analysis methods typically seen by most FDA reviewers. Particularly in light of the evolving nature of scientific knowledge and technical capacity relating to nanotechnology, it will be important to ensure ongoing training, as well as dissemination of new information within review centers and more broadly across the agency, to ensure timely, informed consideration of the most current science.

Products Subject to Premarket Authorization

As discussed more fully below in the Regulatory Policy Issues section, for products subject to premarket authorization, such as drugs, devices, biologics, and food and color additives, FDA reviewers can require manufacturers to provide the necessary scientific information to support regulatory decisions. The evolving scientific understanding of nanoscale materials would generally be expected to inform assessment of what data are needed. For example,one currently would expect the information needed to assess biological interaction to change as size decreases because, as discussed above, data from larger particles may not necessarily predict behavior for smaller particles. However, a precise size boundary where scientific information based on larger particles becomes less relevant for a particular question or material is not currently apparent. Similarly, research identifying short and long-term toxicity issues relevant to particular nanoscale materials could aid FDA in determining what, if any, additional testing a manufacturer should conduct. In short, a greater understanding of the properties of nanoscale materials and of trends in material development and uses would help focus data and testing requirements.

Products Not Subject to Premarket Authorization

For products not subject to premarket authorization by FDA, such as cosmetics and dietary supplements, the agency generally does not receive data, including safety data, before the products are marketed. Furthermore, there are no post-marketing reporting requirements for adverse events associated with cosmetics. Therefore, FDA receives only cosmetic adverse event reports that are submitted voluntarily. Similarly, for dietary supplements, to date FDA has received only voluntarily submitted adverse events, although recently enacted legislation will give FDA greater access to dietary supplement adverse event reports in the future.[24]

The agency may have far less baseline data than with products subject to premarket authorization. As a result, the agency may have a comparatively difficult burden in assembling the necessary data to support a product removal action under these authorities (whether the product contains nanosized or other materials).

Also as discussed above, there may be general differences in properties relevant to evaluation of safety and effectiveness (as applicable) of products using nanoscale materials compared to products using other materials. For example, size, shape, and charge of a nanoscale material can affect disposition or toxicity in the body in ways that differ from molecular forms of materials and that may be generalizable across different particle or other material types. Knowledge of such generalized differences could, for example, help inform FDA's: assessments of whether to take regulatory actions against products not subject to premarket authorization; efforts to obtain and develop further information; and efforts to develop guidance on data needs for products not subject to premarket authorization.

Recommendations for Consideration

The Task Force recommends strengthening FDA's promotion of, and participation in, research and other efforts to increase scientific understanding, to facilitate assessment of data needs for regulated products. Such activities should, where appropriate, be coordinated with and leveraged against activities supported by other Federal agencies, the private sector, or other countries. This would include:

- Promoting efforts, and participating in collaborative efforts, to further understanding of biological interactions of nanoscale materials, including, as appropriate the development of data to assess likelihood of long term health effects from exposure to specific nanoscale materials;
- Assessing data on general particle interactions with biological systems and on specific particles of concern to FDA;
- Promoting and participating in collaborative efforts, to further understanding of the science of novel properties that might contribute to toxicity, such as surface area or surface charge;

- Promoting and participating in collaborative efforts to further understanding of measurement and detection methods for nanoscale materials;
- Collecting/collating/interpreting scientific information, including use of data calls for specific product review categories (see Regulatory Policy Issues section);
- Building in-house expertise;
- Building infrastructure to share and leverage knowledge internally and externally, seeking to collect, synthesize, and build upon information from individual studies of nanoscale materials; and
- Ensuring consistent transfer and application of relevant knowledge through establishment of an agency-wide regulatory science coordination function for products containing nanoscale materials.

Issue: Adequacy of Testing Approaches for Assessing Safety and Quality of Products Containing Nanoscale Materials

Comments

Several comments expressed the concern that existing toxicology screening methods will not adequately assess toxicologic properties of nanoscale materials, and that these methods cannot be used in their present form to assess engineered nanoscale materials. Some comments pointed out that pharmacokinetics and pharmacodynamics of nanoscale particles are different from those of larger particles and that existing toxicity screening studies do not take these differences into account. Several comments further recommended that cellular assays should reflect exposure media (e.g., air, water, food), route of exposure, and portal-of-entry toxicity (e.g., toxicity to lungs, skin, mucus membranes), as well as systemic responses.

Comments stated that most toxicology tests are short-term, and might leave long-term effects unevaluated, especially because the long-term toxicity and effects for most nanoscale materials remain unknown. These comments noted that appropriate endpoints for *in vitro* assays can be difficult to determine, as single cell types are often not sufficient for evaluation of the function or health of organs or tissues that are made up of multiple cell types, and given that various types of tissues are exposed in the body.

Other comments were directed to the inadequacy or absence of currently available and standardized product characterization methodologies for nanoscale materials. A comment highlighted that FDA has limited ability to detect nanoscale material components in some products. Another comment emphasized that FDA's ability to inspect products is also presently significantly limited with regard to products that may contain nanoscale materials. One comment recommended that nanoscale material be characterized with respect to size (surface area and size distribution), chemical composition (such as purity and crystallinity), surface structure (surface reactivity, surface groups, inorganic/organic coatings, etc.), solubility, shape and aggregation. Other comments noted that detection of nanoscale materials requires expensive and sophisticated equipment, and it is often unclear which parameters are relevant to toxicity.

Comments also encouraged FDA to work with other government agencies to develop rapid screening tools for these types of products. Several comments noted that the Nanotechnology Characterization Laboratory, run by the National Cancer Institute, will be very useful in helping to characterize nanoscale materials and to develop standards and standardized methods for measuring nanoscale materials.

Analysis

Assays to Develop Biological Effects Data

Established *in vitro* and *in vivo* assays and predictive models are available to evaluate a variety of endpoints relevant to the establishment of hazard(s) and the identification of further testing needs. The results of these tests are often used in a tiered approach in an overall data development process for understanding the toxicity and effectiveness of a product such that, for example, a "positive" response may lead to a conclusion of hazard or the initiation of additional studies and a "negative" response would not (the obverse may also be true). However, because many of these tests were developed for molecular forms of materials, and nanoscale materials may behave differently, the ability of these tests to support decisions about biological effects or further testing requirements need to be evaluated.

For example, data to support understanding of dose-response developed through *in vitro* test systems might not be appropriate for particles where sedimentation velocities and diffusion can change the delivered dose.[25] Similarly, in some cases it may be necessary to develop information to evaluate whether current short-term tests provide sufficient predictive value regarding the need for chronic or other long-term toxicity testing, and in some cases the only way to get this information may be to actually conduct long-term toxicity testing. In addition, development of new testing methods may be necessary to develop data to support decisions for nanoscale materials that may have novel biological responses.

Existing information for nanoscale materials does not appear to indicate a need for revision to all tests however. Accordingly, a tiered or staged approach to evaluation would seem appropriate. The first stage of such an approach would be to determine whether any specific tests may need evaluation. Subsequent stages would call for data to assess individual test methods as needed.

Standard approaches for handling of materials for testing will also need to be evaluated and may need to be modified, with respect to such factors as appropriate solvents and dosing formulations, methods to prevent agglomeration of particles, stability conditions, and other variables that may affect test results when nanoscale materials are tested. Such adjustments may be needed to improve the reliability, repeatability, reproducibility and accuracy of assays and methods.

Metrology - Characterization of Particles in the Nanoscale Range

Currently, ability to detect nanoscale materials in the body or in products regulated by FDA is limited, and development of appropriate analytical methods for classes of products and of nanoscale materials may require substantial effort. Further, new analytical methods, and methods that FDA reviewers are generally less familiar with, are often used to characterize nanoscale materials. The strengths and limitations of these methods may vary in ways relevant to evaluating characteristics such as particle size, size distribution, surface charge, surface properties, and particle interactions (such as aggregation) that may be relevant to dose, stability, or other characteristics significant to biological interaction or product quality.

Inspection

Tests used for inspections and product surveillance will need to be evaluated to determine whether modifications are warranted to address nanoscale materials. Increased use of materials in the nanoscale range may present particular challenges, for example, relating to tests that assess product stability or development of potentially hazardous byproducts. Further, as with any product, scaling up to full production rates may affect such factors as purity, particle behavior, size distribution, and general batch-to-batch consistency, and it may be necessary to evaluate the adequacy of existing testing to assess such consequences of scale-up for products using nanotechnology.

Standardization of Tests and Data Reporting

In many cases, methods currently used to characterize nanoscale materials have not been standardized through recognized standard development organizations. There is a need to develop standard particle characterization methods. Furthermore, given the range of methods being used and being developed for nanoscale materials, there is a need to develop consistent nomenclature and measurement types and formats for use across studies and data submissions similar to the "MIAME" approach developed for microarray data.[26] Consistent reporting approaches will make data from one laboratory or for one material type more readily comparable to data from other laboratories and more useful to subsequent consideration of properties of the same type of material or other material types.

Recommendations for Consideration

To be marketed, FDA regulated products must be safe and, as applicable, effective. FDA-regulated products must also meet all applicable good manufacturing practice and quality requirements. Adequate testing methods are needed regardless of whether a product is subject to premarket authorization or not. Accordingly, the following recommendations are relevant to all categories of FDA-regulated products. The agency should:

- Evaluate the adequacy of current testing approaches to assess safety, effectiveness, and quality of products that use nanoscale materials;
- Promote and participate in the development of characterization methods and standards for nanoscale materials; and
- Promote and participate in the development of models for the behavior of nanoscale particles *in-vitro* and *in-vivo*.

The Task Force recommends encouraging manufacturers to consult with the agency regarding the appropriateness of testing methodologies for evaluating products using nanoscale materials.

REGULATORY POLICY ISSUES

Background

FDA regulates a broad range of products under the Federal Food, Drug, and Cosmetic Act (FFDCA) and the Public Health Service Act (PHS Act). The agency's statutory authorities subject some types of products to premarket authorization requirements, either individually or by category, while permitting other products to be marketed without prior agency authorization.[27]

Products subject to premarket authorization include drugs, biological products, devices, and food and color additives. As detailed below, new drugs,[28] biological products, and devices receive marketing authorization on a product-by-product basis. FDA authorizes food additives and color additives for marketing by issuing a regulation approving a substance that meets prescribed identity and quality specifications for specified food additive or color additive uses under specified conditions. Once such a regulation is in place, it covers all products that comply with the regulation, and individual premarket review of such products is not required. Drug ingredients can similarly be reviewed for inclusion in monographs authorizing their marketing over the counter (OTC monographs) as generally recognized as safe and effective (GRAS/E) for specified conditions of use. Once the monograph is in place, it covers all products that comply with it, and individual premarket review of such products is not required. FDA-regulated products not subject to premarket authorization include dietary supplements, cosmetics, and food ingredients that are generally recognized as safe (GRAS). Although premarket authorization does not apply, premarket notification is required in some cases for dietary supplements containing new dietary ingredients,[29] and premarket notifications may be submitted at the option of the manufacturer for GRAS food ingredients.[30]

In all cases, whether subject to premarket authorization or not, FDA-regulated products cannot be marketed unless they satisfy specified statutory requirements. In addition to other such requirements, drugs, biological products and devices must be safe and effective; and foods (including dietary supplements and food additives), color additives, and cosmetics must be safe.[31]

As discussed in the State of the Science section, the Task Force believes that nanoscale materials will present regulatory challenges that are similar to those posed by other new technologies FDA has dealt with in the past, such as biotechnology products, but also some potentially new challenges. The Task Force began its regulatory policy inquiry by reviewing the agency's authorities to meet any unique challenges that may be presented by FDA-regulated products containing nanoscale materials. Although FDA's authorities may be adequate to meet these challenges, in some cases the evolving state of the science regarding nanotechnology may warrant a case-by-case approach to assess whether sufficient evidence exists to show that products satisfy the applicable statutory and regulatory standards.

The Task Force has made recommendations that seek to address the challenges nanoscale materials may present. The guidances the Task Force is recommending would give affected manufacturers and other interested parties timely information about FDA's expectations, so as to foster predictability in the agency's regulatory processes, thereby enabling innovation and enhancing transparency, while protecting the public health.

The very nature of nanoscale materials – their dynamic quality as the size of nanoscale features change, for example, and their potential for diverse applications – may permit the development of highly integrated combinations of drugs, biological products, and/or devices, having multiple types of uses, such as combined diagnostic and therapeutic intended uses. As a consequence, the adequacy of the current paradigm for selecting regulatory pathways for "combination products"[32] may need to be assessed to ensure predictable determinations of the most appropriate pathway for such highly integrated combination products.

Continuing agency efforts to gather together and share scientific knowledge of nanoscale materials, across centers and divisions, will be important to facilitate informed regulatory activity in light of the evolving state of this scientific knowledge. Such coordinated assessment and dissemination of information will enable informed determinations of how best to approach premarket review, including identification of appropriate regulatory pathways for such highly integrated combination products. Among other uses, it will also support assessment of postmarket surveillance of products using nanoscale materials.

As the recommendations below reflect, the Task Force believes communication between regulated entities and the agency early in the product development process, particularly with regard to highly integrated combination products, will help ensure timely consideration of any potentially novel issues that products using nanoscale materials may raise. In addition, to assist the agency to be well-positioned to enable the development and premarket review of such highly integrated combination products, the Task

Force recommends that FDA seek public input on the adequacy of agency policies and procedures.

Discussion

The Task Force has considered four broad regulatory policy questions that concern FDA in relation to the presence of nanoscale materials in FDA-regulated products. These questions are:

- Is FDA able to determine whether particle sizes or material features of products extend into the nanoscale range?
- What is the scope of the agency's authority regarding the evaluation of the safety and effectiveness of products containing nanoscale materials?
- Should product labeling declaring the presence or amount of nanoscale materials be either required or permitted?
- Does the use of nanoscale materials in FDA-regulated products raise any issues under the National Environmental Policy Act (NEPA)?

After summarizing the agency's general authorities, we address each question in turn below, taking into account relevant comments submitted to the public docket for the FDA Nanotechnology Task Force's October 2006 public meeting and to the docket opened in response to the ICTA Petition.

Summary of Agency Oversight Authorities by Product Type

Following are concise summaries by product type of certain agency statutory and regulatory authorities relating to product safety, effectiveness (as applicable), and quality. This discussion is relevant, in particular, to the Task Force's analyses and recommendations with respect to the first and second questions presented above.

New Drugs and Biological Drug Products

Sponsors are required to submit an Investigational New Drug (IND) application to FDA in accordance with 21 CFR Part 312 prior to conducting human clinical studies of most drugs.[33] INDs are required to contain detailed information about the investigational new drug, including chemistry, manufacturing, and controls information, such as information about its active ingredients and structural formula, and pharmacological and toxicological results from studies of the drug in animals. During FDA's review of the data contained in the IND submission, the agency may identify additional information necessary to assure the safety of subjects and assure that the study design is adequate to permit an evaluation of the drug's safety or effectiveness in humans. FDA has the authority to request such additional safety data from the sponsor, including particle size data, when it is needed to support the IND.

After a drug has been adequately studied in humans, the applicant must submit a new drug application (NDA) to obtain approval to market the drug (21 CFR § 314.50). For biological products regulated under section 351 of the PHS Act, an applicant must submit a biologics license application (BLA) (21 CFR § 601.2). Applicants are required to submit in NDAs and BLAs detailed technical information about their products, including chemistry, manufacturing, and controls information, the results of animal pharmacology and toxicology studies, bioavailability data, and extensive data on safety and effectiveness generated in clinical investigations of the drugs in humans.

Applicants may also seek FDA approval of an abbreviated new drug application (ANDA) to market a generic version of a drug for which the agency has previously approved an NDA. ANDAs contain detailed technical information about the product, including chemistry, manufacturing, and controls information, and bioavailability and bioequivalence data, but do not include extensive human data from clinical investigations (21 CFR § 314.94).

During FDA's review of these applications, the agency may call for additional data from the sponsor needed to support the applications, including particle size data, if not supplied in the original application. FDA requests information on particle size when the agency considers such information relevant to determining whether a particular human drug product or class of human drug products is safe and effective. If FDA determines such data are needed for a class of drugs, FDA may issue guidance to applicants recommending that they be submitted in the original application.

Drugs, including biological drugs, are also subject to current good manufacturing practice requirements found in 21 CFR Parts 210 and 211.

These requirements govern the methods to be used in, and the facilities or controls to be used for, the manufacture, processing, packing, or holding of a drug. They are intended to ensure that the drug meets the safety requirements of the FFDCA, and has the identity and strength and meets the quality and purity characteristics that it purports or is represented to possess.

In accordance with 21 CFR § 314.70, 21 CFR § 314.97, and 21 CFR § 601.12, after a human drug or biologic product is approved for marketing, the sponsor must notify FDA about each change in each condition established in the approved application. Significant changes require a supplement to the NDA, ANDA, or BLA, and prior FDA approval. Significant changes include a change in drug substance, drug product, production process, quality controls, equipment or facilities that has a substantial potential to have an adverse effect on the drug product. Changes to a product to introduce nanoscale ingredients or processing would trigger change notification chemistry supplements and permit FDA to review and approve the revised formulation. Depending on the change, the resulting product might be considered a new product for which a new approval is needed.

OTC Monograph Drugs

Over-the-counter drug monographs establish the active ingredients and conditions of use for OTC drugs that qualify as generally recognized as safe and effective. Additional active ingredients can be added to an OTC drug monograph in response to a citizen petition (under 21 CFR § 10.30) or a time and extent application for eligibility (under 21 CFR § 330.14) followed by evaluation of data supporting GRAS/E status. Both processes require the submission of safety and effectiveness data. FDA can require data and information to determine if these proposed additional ingredients contain nanoscale materials and, if so, require safety and effectiveness data directly related to particle size to determine whether the ingredient qualifies for inclusion in the monograph.

If a manufacturer does not have an approved NDA for a drug product that has not been marketed in the United States, that manufacturer generally cannot market that drug product unless FDA has published a final OTC monograph that includes the drug product's active ingredient for the intended use.[34] Products containing active ingredients that are already included in a monograph and that bear labeling published in a monograph may be marketed without product-specific premarket authorization. FDA can take various actions however, if the agency learns that a new version of a drug product marketed under an OTC monograph raises a safety or effectiveness concern. A new version that might raise such concerns could be a drug product that contains a monograph ingredient whose particle size has been reduced to the nanoscale range To address this situation, for example, the agency can issue a call for data on the safety and effectiveness of the changed version of the OTC drug ingredient. In addition, under 21 CFR Part 330, FDA can conduct rulemaking to determine whether a nanoscale version of a monograph ingredient should be considered nonmonograph (i.e., not GRAS/E), and therefore to require submission of data in an NDA to establish its safety and effectiveness.

New Animal Drugs and Animal Feed Containing a New Animal Drug

New animal drugs, including new animal drugs for use in animal feed, are regulated under section 512 of the FFDCA and are subject to a premarket authorization process intended to establish the products' safety and effectiveness that is in many respects similar to the process used for new human drugs and biological drug products. An applicant must submit a new animal drug application (NADA) to obtain approval to market the animal drug and the requirements for the contents of NADAs are found in 21 CFR Part 514. Animal drugs must also be manufactured in accordance with current good manufacturing practice requirements found in 21 CFR Part 211. FDA has the authority to require information that will adequately characterize the drug formulation, identify quality-indicating specifications, and ensure that factors affecting the quality, purity, strength and potency are adequately understood and controlled. This authority includes the ability to require the submission of a product's particle size, where the particle size might have an impact on the safety or effectiveness of the animal drug.

Devices

Medical devices are regulated under the FFDCA according to a tiered classification system that is largely based on the degree of risk posed by the product. Devices that are low risk, for which safety and effectiveness are generally well-established, are designated as Class I devices. These device types are subject to general controls, such as labeling, good manufacturing practices and adverse event reporting.

Class II devices are more complex, and carry a higher risk than Class I devices. Before marketing the product, manufacturers are usually required to submit a premarket notification under section 510(k) of the FFDCA (510(k) submission) for FDA review. In a 510(k) submission, manufacturers are required to submit data and other information to demonstrate that their device is "substantially equivalent" with regard to safety and effectiveness to a similar device already legally marketed in the United States. The evidence provided usually consists of pre-clinical testing (such as animal, bench, and analytical testing), and occasionally, clinical data (such as data derived from a study using a patient population with a defined clinical condition).

In some cases, manufacturers may make modifications to Class I or Class II devices after FDA clearance without submitting a new 510(k) submission. FDA provides guidance on when a new 510(k) is needed for a modified device in the guidance document entitled, "Deciding When to Submit a 510(k) for a Change to an Existing Device (K97-1)."[35] Manufacturers of a commercially distributed Class I or Class II device, for which FDA has granted an exemption from the requirement of a 510(k) submission for the generic type of device, must still make such a submission under certain circumstances, such as when the modified device operates using a different fundamental scientific technology than a legally marketed device in that generic type of device. Accordingly, manufacturers would have to make a submission if use of a nanoscale material were to qualify as a use of a different fundamental scientific technology.

Class III devices are the most complex, high risk devices. These devices are reviewed under a premarket approval application (PMA). In a PMA, manufacturers provide detailed evidence that their device provides a "reasonable assurance of safety and effectiveness." The evidence provided usually consists of pre-clinical testing, and clinical data. Class III PMA devices are subject to a pre-approval manufacturing inspection and require submission of periodic reports. Any modification to a PMA device and its method of manufacture that could affect safety or effectiveness requires approval by FDA through the submission of a PMA supplement. A PMA might be required for a product otherwise within a general category considered Class I or Class II if the inclusion of nanoscale material raises questions of safety or effectiveness warranting clinical studies.

Manufacturers are required to submit information in sufficient detail to describe the device and its intended use. This includes, for example, information on the chemical composition and physical characteristics of materials that comprise the device. In general, devices must be manufactured in accordance with the Quality System Regulation (21 CFR Part 820).

All clinical evaluations of investigational devices, unless exempt, must have an approved investigational device exemption (IDE) before the study is initiated. An approved IDE is required for "significant risk devices" but not for "nonsignificant risk" devices. A significant risk device presents a potential for serious risk to the health, safety, or welfare of a subject.

Significant risk devices may include implants, devices that support or sustain human life, and devices that are substantially important in diagnosing, curing, mitigating or treating disease or in preventing impairment to human health. Examples include sutures, cardiac pacemakers, hydrocephalus shunts, and orthopedic implants. Clinical studies of devices that pose a significant risk require both FDA and Institutional Review Board (IRB) approval prior to initiation of the study. FDA approval is obtained by submitting an IDE application to FDA (21 CFR § 812.20). If use of a nanoscale material were to pose a potential for serious risk to health, safety, or welfare of a subject, an IDE would, therefore, be required.

Nonsignificant risk devices are devices that do not pose a significant risk to human subjects. Examples include ultrasonic dental scalers, conventional laparoscopes, culdoscopes, hysteroscopes, and foley catheters. A nonsignificant risk device study requires only IRB approval prior to initiation of a clinical study, and sponsors are not required to submit an IDE application to FDA for approval. If the IRB disagrees with the sponsor and determines that the device poses a significant risk, the sponsor must report this finding to FDA.

Guidance on distinguishing between significant risk and nonsignificant risk studies is outlined in the FDA guidance document entitled "Information Sheet Guidance for IRBs, Clinical Investigators, and Sponsors, Significant Risk and Nonsignificant Risk Medical Device Studies."[36]

Food Additives and GRAS Food Ingredients

FDA regulates food additives and generally recognized as safe (GRAS) food ingredients under sections 201(s) and 409 of the FFDCA. Under these statutory provisions, any substance added to food "directly or indirectly" is a food additive unless the substance is GRAS for its intended use, is a pesticide, or is otherwise excluded from the definition of a food additive. Food additives must receive premarket approval from FDA in the form of a regulation establishing conditions of safe use. Food additives include those substances added directly to food, substances that may become components of food as a result of their use in processing, and components of food contact materials that can reasonably be expected to migrate to food.

Section 409 of the FFDCA and 21 CFR Parts 170 and 171 describe in general terms the information and data necessary to establish the safety of food additives and ingredients. These authorities are supplemented by technical guidance documents providing more specific data recommendations. In addition, FDA may require any other information that it determines during its review is necessary to establish safety.

The specific data that FDA can require to establish the safety of food additives include information on: identity, including physical characteristics such as particle size; the physical or chemical technical effect of the additive; analytical methods for determining the quantity of the substance in food; and the safety of the intended use of the substance. These requirements exist regardless of the physical or chemical characteristics or physical state of the additive. Where appropriate to ensure safety, FDA places limitations on the physical and chemical properties of food additives, which include particle size.

For an approved food additive, FDA publishes a final regulation establishing conditions under which the substance may be safely used. Like an OTC monograph, this rule applies to products that satisfy these conditions. FDA can take various actions however, if the agency learns that a new version of a substance being marketed under a food additive regulation raises safety concerns. A new version that might raise such concerns could be a food additive that contains or may contain nanoscale materials. In such a situation, for example, the agency can issue a call for data on the safety of such a version of the substance. In addition, under 21 CFR Parts 170 and 171, FDA can publish a proposed rule to amend the food additive regulation to address under what circumstances the nanoscale version of the substance may be safely used.

GRAS uses of food ingredients do not require premarket authorization by FDA. Nonetheless, the safety data and information to support GRAS uses of food ingredients must be of the same quality and quantity as data needed to establish the safety of a food additive. In addition, the data must be generally available. Finally, for uses of food ingredients to be GRAS, the safety of the ingredient must be generally recognized by scientists qualified to assess the safety of such substances. As part of its GRAS notice process, FDA can inform manufacturers of what data the agency considers necessary to establish the safety of food ingredients.

Color Additives

FDA regulates color additives under sections 201(t) and 721 of the FFDCA. Generally under these statutory provisions, any substance capable of imparting color to any food, drug, cosmetic, or medical device, or the human body is a color additive that requires premarket approval by FDA in the form of a regulation listing (i.e., approving) the color additive for its intended uses. In addition to being used in compliance with an existing regulation, some color additives may require batch certification by FDA. During the premarket approval process for color additives, FDA reviews detailed manufacturing and analytical data to judge whether postmarket batch certification will be required. In many cases this part of the premarket authorization process requires the sponsor to provide batch samples of the color for analysis by FDA color chemists. The judgment regarding whether a color requires batch certification is ordinarily based on the expected variation in manufactured color and on the degree of concern that such variation raises regarding the safety of the color additive.

Section 721 of the FFDCA and 21 CFR Part 71 describe in general terms the information and data necessary to establish the safety of color additives. These authorities are supplemented by technical guidance documents providing more specific data recommendations. In addition, FDA can generally require the submission of any data that it determines in its review to be necessary to establish safety. The specific data that FDA can require to establish the safety of a color additive include information on: identity, including physical characteristics such as particle size; analytical methods for determining the quantity of the substance in the finished product and for ensuring the purity and consistency of the manufactured color; and the safety of the color additive under its intended conditions of use. These requirements exist regardless of the physical or chemical characteristics or physical state of the color additive. Where appropriate to ensure safety, FDA places limitations on the physical and chemical properties of color additives, which include particle size.

Once FDA has promulgated a regulation listing a color additive, the regulation applies to products that comply with the rule's conditions. However, FDA can take various actions if the agency learns that a new version of a substance being marketed under a color additive regulation raises safety concerns. A new version that might raise such concerns could be a color additive that contains or may contain nanoscale materials.[37] In such a situation, for example, the agency can issue a call for data on the safety of such a version of the substance. In addition, under 21 CFR Part 71, FDA can publish a proposed rule to amend the listing regulation to address under what circumstances the nanoscale version of the substance may be safely used.

Cosmetics

Section 201(i) of the FFDCA defines a cosmetic as an article intended to be rubbed, poured, sprinkled, or sprayed on, introduced into, or otherwise applied to the human body or any part thereof for cleansing, beautifying, promoting attractiveness, or altering the appearance, and articles intended for use as a component of any such articles; except that such term shall not include soap.[38] This definition includes skin-care creams, lotions, hairsprays, perfumes, lipsticks, fingernail polishes, eye and facial makeup, permanent waves, hair colors, deodorants, baby products (e.g., baby powder, baby oil, wipes), bath oils, bubble baths, and mouthwashes, as well as any material intended for use as a component of a cosmetic product. Under the FFDCA, FDA is not given premarket approval authority for cosmetic products and most cosmetic ingredients (other than color additives). However, FDA's mission includes ensuring that cosmetics are safe and properly labeled. FDA pursues this mission through a combination of activities which include inspection of cosmetic manufacturing establishments and enforcement actions for cosmetic products found to be in violation of the "adulteration" (Section 601) and/or "misbranding" (Section 602) provisions of the FFDCA. For example, if a cosmetic bears or contains any poisonous or deleterious substance which may render it injurious to users when used under labeled or customary conditions of use, the product is adulterated.

Dietary Supplements

Dietary supplements are regulated under the FFDCA, as amended by the Dietary Supplement Health and Education Act of 1994, or DSHEA. In addition, there are implementing regulations in 21 CFR Parts 101, 119, and

190. As defined in the statute (21 U.S.C. 321(ff)), a dietary supplement is a product other than tobacco that is intended to supplement the diet and that contains one or more dietary ingredients. A dietary ingredient is any one of the following: a vitamin; mineral; herb or other botanical; amino acid; dietary substance for use by man to supplement the diet by increasing the total daily intake; or a concentrate, metabolite, constituent, extract or combination of any of the other types of dietary ingredients. The requirement to contain a dietary ingredient is only one element of the dietary supplement definition. A dietary supplement must also be labeled as such and must be intended for ingestion. Topical products like lotions and ointments cannot be dietary supplements. A dietary supplement must not be represented for use as a conventional food or as a sole item of a meal or the diet. Generally, articles approved as new drugs, licensed as biologics, or authorized for clinical investigation under an IND cannot be marketed as dietary supplements. However, if the product was marketed as a dietary supplement or as a food before such approval, licensing, or authorization under an IND, it may still be marketed as a dietary supplement afterwards.

With one exception, FDA has no authority to require premarket safety testing or premarket submission of safety information for a dietary supplement. The manufacturer of a dietary supplement must notify FDA at least 75 days in advance of marketing a product if it contains a new dietary ingredient,[39] unless that ingredient has been "present in the food supply as an article used for food in a form in which the food has not been chemically altered" (21 U.S.C. 350(b)). The manufacturer must include in the notification the information on which the manufacturer based its conclusion that a dietary supplement containing the new dietary ingredient will reasonably be expected to be safe. However, the nature of the safety information on which the manufacturer may rely is not specified in the law, and there is no requirement that a manufacturer wait for a safety determination from FDA before marketing the product.

Because the majority of dietary supplements do not contain a new dietary ingredient, most dietary supplement safety issues arise in the post-market context. As with conventional foods and cosmetics, a manufacturer may market a dietary supplement without pre-market authorization from FDA. As part of its responsibility to ensure that FDA-regulated products sold in the U.S. are safe and properly labeled, the agency inspects dietary supplement manufacturing facilities and takes action against dietary supplements that are adulterated (21 U.S.C. 342) or misbranded (21 U.S.C. 343).

Because FDA's regulation of dietary supplements is generally post-market, the agency may not know whether particle sizes or material features used in dietary supplement products are in the nanoscale range, unless the agency becomes aware of the use of such sizes and features, for example, from information submitted in a notification or from the product labeling.

Labeling and Advertising Claims for Conventional Foods and Dietary Supplements

In some cases, FDA may become aware of the use of nanoscale materials or features in a dietary supplement or conventional food because of labeling or advertising claims for the product in the marketplace. These could be claims that explicitly reference the presence of nanoscale materials in the product or claims that describe product characteristics or effects derived from the use of nanoscale materials or nanotechnology.

In addition, there are some labeling claims for conventional foods and dietary supplements that require a premarket petition process or a postmarket notification to FDA. For example, health claims (i.e., claims about reducing the risk of a disease or health-related condition) for a conventional food or dietary supplement require premarket review by FDA. The mechanism for requesting a new health claim is a health claim petition under 21 CFR 101.70, requesting that the agency issue a regulation regarding the claim. Health claim petitions must explain the petitioner's basis for concluding that the substance that is the subject of the requested claim is safe, and petitions must also include studies demonstrating that the substance has the claimed risk reduction effect.

For dietary supplements, the FFDCA (Section 343(r)(6)) specifically authorizes the use of labeling claims concerning effects on the structure or function of the body (structure/function claims), claims of general well-being, and claims of a benefit related to a classical nutrient deficiency disease. These claims do not require premarket review, but they do require notification to FDA within 30 days of marketing. Unlike a health claim petition, however, a postmarket notification for a dietary supplement labeling claim under Section 403(r)(6) of the Act need not include safety information about the substance for which the claim is being made or studies demonstrating that it has the effect claimed.

In summary, FDA review of a labeling claim, such as a structure or function claim or health claim, would not necessarily include information related to use of nanoscale materials. Whether the agency would have the opportunity to review such information would depend on whether the nanoscale material was related to the basis for the labeling claim and, if so, whether the information about the nanoscale material was submitted to the agency, e.g., in a health claim petition as required by 21 CFR § 101.70(f), or voluntarily for other types of claims.

ISSUE: ABILITY OF FDA TO IDENTIFY FDA-REGULATED PRODUCTS THAT CONTAIN NANOSCALE MATERIALS

Comments

Several comments stressed the importance of identifying the use of nanoscale materials in FDA-regulated products. One comment suggested that companies generally submit information on particle size. Another comment urged FDA to call for data on uses of nanotechnology for foods. Another comment suggested that FDA establish a mandatory premarket notification system for novel uses of nanotechnology and urged FDA to issue guidance defining "novel use." Several comments discussed the importance of a clear definition of nanoscale materials in order to eliminate confusion. A few comments stated it was necessary to distinguish between "common" nano-sized food ingredients and those intentionally nano-sized.

Analysis

As discussed in the State of the Science section, although the science of nanotechnology is continuing to evolve, it is known that the size of a particle can affect its properties such that versions of the same substance with differing particle sizes can have different properties, such as their interaction with light (e.g., transparency of lotions). To appropriately assess the safety and, as applicable, the effectiveness of products, it will be important in some cases for FDA, or the manufacturer, to take into account whether the product contains nanoscale materials. FDA's authority to obtain information on particle size differs depending on whether products are subject to premarket authorization or not. As indicated above, the agency's authority is comprehensive with regard to products subject to premarket authorization such as drugs, devices, biological products, and food and color additives. FDA's authority is more limited with regard to products that are not subject to premarket authorization, which include cosmetics and dietary supplements.

FDA can require submission of data regarding particle size and other relevant properties when such data are necessary to evaluate the safety or, as appropriate, effectiveness of a product that is subject to premarket authorization. However, the agency might not obtain such information if the manufacturer is uncertain of whether it needs to obtain authorization to market its product or already has authorization to do so (e.g., of whether its product is covered by an existing food or color additive regulation or by an OTC drug monograph). When dealing with products not subject to premarket authorization, the agency has less ability to obtain information about the presence of nanoscale materials.

Recommendations for Consideration

Issue guidance to sponsors regarding identification of the particle size for:

- Products subject to premarket authorization, including OTC drugs (when a new monograph or amendment to a monograph is being proposed), and food and color additives (in petitions to approve new additives or to amend existing approvals); and
- Products not subject to premarket authorization but for which the sponsor is required to provide notice (such as dietary supplements containing certain new dietary ingredients), or may choose to provide notice (such as a GRAS notification).

When warranted, issue a call for data to identify:

- OTC drug products that contain or may contain nanoscale versions of ingredients included in an OTC monograph; and
- Nanoscale versions of previously approved food and color additives.

Issue: Scope of FDA's Authority Regarding Evaluation of Safety and Effectiveness

Comments

Several comments requested that FDA collect safety information on the use of nanotechnology in regulated products. Most comments stated that there should be disclosure, transparency, and sharing of scientific information. One comment suggested that all existing safety data should be submitted. One comment recommended that FDA request available studies on nano- and micro-scale ingredients to substantiate safety in personal care products. One comment urged cosmetic manufacturers to submit substantiation data voluntarily. Other comments urged obligatory reporting of safety and health problems caused by nanoscale products.

Opinions differed regarding the need to develop a regulatory framework specific to nanotechnology. Most of the comments urged regulation based upon the unique risks of small particle size. One comment, for example, stated that FDA should start with the assumption that nanoscale materials in products behave in a distinct way and, therefore, should be subject to nanospecific paradigms and health and safety testing. On the other hand, many of the comments stated that there was not enough scientific evidence of unique risks posed by products using nanotechnology and therefore disagreed that FDA should establish a distinct regulatory regime for products using nanotechnology. These comments stated that the regulatory pathways currently utilized by FDA based on statutory classification of the products are sufficient to ensure the safety and effectiveness of products using nanotechnology. These comments urged FDA to evaluate products according to consistent, proven safety standards. Still other comments stated the need to address "intentionally produced" nanoscale materials. One comment stressed the need to avoid definitions that failed to distinguish between common, naturally occurring nanoscale food components and those intentionally used components that might, based upon the hazards posed, require modification of existing regulatory frameworks.

Several comments stated that products using nanotechnology should be treated as new products. Some groups wanted all products using nanotechnology to be subjected to safety testing prior to marketing. One comment suggested that use of nanoscale materials could change the regulatory pathway used to ensure safety and that FDA establish criteria for what is "new for legal and regulatory purposes" and "new for safety evaluation purposes."

Some comments focused on the safety standard for products using nanotechnology. One comment suggested that the agency adopt a "reasonable certainty of no harm" standard. Several comments requested guidance on substantiation, one of which recommended that substantiation be based on the "competent and reliable scientific evidence" standard under the Federal Trade Commission Act.

Several comments focused on the process for developing timely and responsible regulations. These comments urged that any regulation to identify and minimize risks should be adopted in an open and transparent process.

One comment stated that FDA has ample legal authority to require food manufacturers to establish safety, substantiate label claims based on sound science, and remove unsafe products from the market.

Analysis

Products Subject to Premarket Authorization

Because nanoscale materials can behave differently than other versions of the same materials, it will be important for FDA to obtain relevant information about the characteristics of products containing nanoscale materials. The guidance recommended in the section above entitled "Ability of FDA to Identify FDA-Regulated Products that Contain Nanoscale Materials" would assist the agency in identifying the particle size and other relevant properties of nanoscale materials in products subject to FDA's review. As stated above, FDA's authority over products subject to premarket authorization is comprehensive and provides FDA with the ability to obtain detailed scientific information needed to assess the safety and, as applicable, effectiveness of products, including relevant effects of nanoscale materials. In some cases, the presence of nanoscale materials may change the regulatory status/regulatory pathway of products. The Task Force believes it is important that manufacturers and sponsors be aware of the issues raised by nanoscale materials and the possible change in the regulatory status/pathway when products contain nanoscale materials.

Recommendations for Consideration

To provide clear guidance to interested parties and to enhance FDA's knowledge base, the Task Force recommends that the agency take the following actions regarding products subject to premarket authorization:

- Issue a notice in the Federal Register requesting submission of data and other information addressing the effects on product safety and effectiveness of nanoscale materials in products subject to FDA premarket authorization, including both existing products that are changed to include (or include greater proportions of) nanoscale materials and new products made with nanoscale materials. In addition, the Task Force recommends that FDA seek comment on whether FDA's current policies for determining the appropriate regulatory pathway are optimal for reaching timely and predictable decisions for highly integrated combination products containing nanoscale materials. If commenters believe current policies are not optimal, they would be asked for examples of the kinds of products for which FDA's policies are not optimal and why they are not.
- Issue guidance requesting submission of information on whether and how the presence of nanoscale materials affects the manufacturing process for products subject to premarket authorization, as part of a premarket submission. Relevant information would address situations when the product contains nanoscale materials and when any part of the manufacturing process involves nanoscale materials, even if those materials do not become part of the finished product.

Issue guidance or amend existing guidance to describe what additional or distinct information should be submitted to FDA or generated with regard to the following:

- New food or color additives made with nanoscale materials; and
- Previously approved food or color additives that are now made with nanoscale materials or to contain greater proportions of nanoscale materials.

Issue guidance describing when:

- A sponsor of a Class I or Class II device, who is otherwise exempt from submitting a 510(k), would need to submit a 510(k) because the presence or amount of nanoscale material would result in the device being outside the scope of the limitations of exemption described in the general provisions of the applicable regulations (see 21 CFR §§ 862.9-892.9);
- A sponsor should submit a new 510(k) for a modification to a previously cleared device that incorporates the use or increased use of nanoscale materials; and
- IRBs, investigators, and industry should seek input from FDA on significant risk/nonsignificant risk decisions regarding investigational devices containing nanoscale materials.

Products Not Subject to Premarket Authorization

Where products are not subject to premarket authorization, manufacturers generally are not required to submit data to FDA prior to marketing, and agency oversight capacity is, therefore, less comprehensive. However, manufacturers are still responsible for ensuring that the products they market are safe. For example, cosmetic manufacturers are required to ensure the safety of their products but are not required to provide safety data to FDA. In light of the evolving state of the science, the Task Force believes an appropriate course of action at this time would be for the agency to work with manufacturers of these products and assist them in identifying data to substantiate the safety of products containing nanoscale materials, including chronic toxicity and other long-term toxicity data as appropriate.

Recommendations for Consideration

We recommend the following regarding products not subject to premarket authorization:

- Issue a notice in the Federal Register requesting submission of data and other information addressing the effects on product safety of nanoscale materials in products not subject to premarket authorization. The notice would address both new products made with nanoscale materials and existing products that are changed to include or include greater proportions of nanoscale materials.

- Issue guidance or amend existing guidance to describe what additional or distinct information should be submitted to FDA or generated with regard to:
    - The use of nanoscale materials in food ingredients for which a GRAS notification is submitted or the reduction of particle size into the nanoscale range for food ingredients for which an earlier notification had been submitted and not objected to by FDA; and
    - The use of nanoscale materials in new dietary ingredients.
- Issue guidance recommending manufacturers consider whether and how the presence of nanoscale materials affects the manufacturing process. Relevant considerations would include both situations when the product contains nanoscale materials and when any part of the manufacturing process involves nanoscale materials, even if those materials do not become part of the finished product.
- Issue guidance describing safety issues that manufacturers should consider to ensure that cosmetics made with nanoscale materials are not adulterated.
- Issue guidance on whether a dietary ingredient modified to include nanoscale materials or include a greater proportion of nanoscale materials would still qualify as a dietary ingredient under 21 U.S.C. 321(ff) (1), and on when the reduction in size into the nanoscale range of an "old" dietary ingredient might trigger the notification process required for a new dietary ingredient on the basis of the presence or amount of nanoscale materials.

Issue: Permissible and Mandatory Labeling

Comments

Several comments urged the disclosure in labeling of the presence of nanoscale materials in FDA-regulated products. A few stressed the importance of clear definitions of "nanoengineered materials and nanotechnology" to enable such disclosure. One group recommended that FDA enforce warning label requirements for cosmetics.

Analysis

Consumers are increasingly exposed to information about nanotechnology products, but they may not always understand whether the use of nanotechnology has a significant effect on the products they purchase. As with many other new technologies, the use of nanotechnology does not mean that a product's safety or effectiveness is necessarily increased, decreased, or affected in any way. As the comments reflect, consumers may have questions regarding the use of nanotechnology for products regulated by FDA. Whether information on such use must be included in product labeling, or can voluntarily be included, depends on whether its inclusion is required or permitted under the FFDCA.

The FFDCA requires that labeling of FDA-regulated products be truthful and not misleading. Labeling must include material information, including with respect to consequences which may result from the use of the product under the conditions of use prescribed in the labeling or under customary or usual conditions of use. The risk information contained in prescription drug labeling is an example of material information. For foods, information about the characteristics of the food can be material, such as nutritional, organoleptic (e.g., taste, smell, or texture), or functional (e.g., storage) properties. If labeling is false or misleading, the product is "misbranded" and cannot be marketed.

For products subject to premarket authorization, the agency generally considers on a case-by-case basis as part of marketing authorization whether labeling contains adequate information to support the safe and, as applicable, effective use of the product. Labeling for products not subject to premarket authorization (for example, cosmetics) also must include material information and not include false or misleading information. Otherwise, these products would be misbranded. If FDA determined that a particular use of a specific nanoscale material, or the use of nanoscale materials more generally, was a material fact for a category of products, FDA could amend its regulations to require, for example, that all members of that category of products include labeling regarding such use of nanoscale material.

If disclosure of information on the use of nanoscale materials is not material (and, therefore, is not required to be included in the product labeling), the manufacturer may still be able to include such information, as long as the information is not false or misleading. However, because claims regarding the use of nanoscale materials might be misleading and, therefore, misbrand a product, the Task Force would recommend encouraging manufacturers to consult with the agency concerning such labeling to avoid misbranding the product.

Recommendations for Consideration

Because the current science does not support a finding that classes of products with nanoscale materials necessarily present greater safety concerns than classes of products without nanoscale materials, the Task Force does not believe there is a basis for saying that, as a general matter, a product containing nanoscale materials must be labeled as such. Therefore the Task Force is not recommending that the agency require such labeling at this time. Instead, the Task Force recommends that the agency take the following action:

- Address on a case-by-case basis whether labeling must or may contain information on the use of nanoscale materials.

Issue: National Environmental Policy Act

Comments

No comments were submitted to the Task Force's docket or to the docket for the ICTA Petition regarding NEPA obligations.

Analysis

NEPA requires federal agencies to assess the environmental impacts of "major federal actions" and to ensure that the interested and affected public is informed of environmental analyses. Agencies may prepare an environmental assessment (EA) or environmental impact statement (EIS), as appropriate. Agencies can also establish categorical exclusions for categories of major federal actions that have been determined not to individually or cumulatively have a significant effect on the human environment. However, agency procedures must also provide for those circumstances where a normally excluded action may have a significant environmental effect. Examples of such "extraordinary circumstances" are identified in FDA's regulations and include actions for which available data establish that there is the potential for serious harm to the environment and actions that adversely affect an endangered, threatened, or otherwise specially protected species.

Many FDA regulatory actions constitute major federal actions, including: actions to approve or withdraw applications to market new drugs and biological products; actions to approve or prohibit or otherwise restrict the use of a substance in food, or food packaging; and actions on premarket notifications and premarket applications for devices. Under FDA's regulations, many of these actions can qualify for a categorical exclusion.

The agency requires applicants and petitioners to submit an EA or a claim of categorical exclusion when requesting agency action. An EA must address the relevant environmental issues and contain sufficient information to enable the agency to determine whether the proposed action may significantly affect the quality of the human environment. Agency guidance provides recommendations for preparation of EAs and the making of categorical exclusion claims. The science related to the environmental implications of nanoscale materials is evolving. In some cases, the presence of nanoscale materials may warrant a closer look at potential environmental impacts of an FDA-regulated product. Such products may not qualify for a categorical exclusion or an extraordinary circumstance may exist.

As knowledge of nanoscale materials increases, it may be productive for the agency to develop or amend agency NEPA guidance to address expressly nanoscale materials or certain types of nanoscale materials. In light of the current, evolving state of scientific knowledge regarding nanoscale materials, however, the Task Force recommends a case-by-case approach at this time to assessing NEPA requirements for products using these materials, and coordination across the agency to enable consistent determinations informed by the most current science.

Recommendations for Consideration

We recommend that the agency take the following actions:

- Take into account, on a case-by-case basis, whether an FDA-regulated product containing nanoscale materials qualifies for an existing categorical exclusion and whether extraordinary circumstances exist.
- Designate a lead in the agency to coordinate the agency's approach to its obligations under NEPA regarding nanotechnology.

End Notes

[1] Among other requirements, the Federal Food, Drug, and Cosmetic Act requires that drugs and devices be safe, and the Public Health Service Act requires that biological products be safe. Among other requirements, the FFDCA states that food and cosmetics must not be

adulterated. For convenience, in this report the term "safe" is used not only in reference to drugs, devices, and biological products but also in reference to foods and cosmetics as an approximate short-hand for "not adulterated." See 21 USC 342, 361, defining the term "adulterated" for foods and cosmetics. In addition to regulating their safety, the agency also regulates use claims made with regard to foods and cosmetics, including claims regarding the effects of these products. However, in keeping with the primary focus of the Task Force's mandate, this report addresses mainly issues relating to the safety of these products.

[2] See for example descriptions of research and development activities at: http://nint-innt.nrc-cnrc.gc.ca/research/index_e.html; *http://www.jst.go.jp/* kisoken/nano/en/ Virlabo Research Y01.htmlhttp://nano.cancer.gov/resource_center/cancer_nanotechnology_brochure.pdf.

[3] For additional information on the Task Force, including its membership, see http: //www.fda.gov/nanotechnology/nano_tf.html.

[4] The Task Force notes that the use of the terms "nanoscale materials," "nanoscale particles", and "nanotechnology" in this report does not reflect a Task Force position as to how these or related terms should be used or defined for regulatory purposes, including for purposes of any information requests or guidance recommended in this report.

[5] Heinze, T. 2004. Nanoscience and nanotechnology in Europe: Analysis of publications and patent applications including comparisons with the United States. *Nanotechnology Law & Business* 1(4): 10.

[6] Australian Safety and Compensation Council. A Review of the Potential Occupational Health & Safety Implications of Nanotechnology. July 2006. ISBN 0 642 32609 6; Borm,P.J.A.,et al. 2006. The potential risks of nanomaterials: a review carried out for ECETOC. *Particle and Fiber Toxicology* 3: 11. National Nanotechnology Initiative (NNI). 2006. Environmental, Health, and Safety Research Needs for Engineered Nanoscale Materials. Available at www.nano.gov.

[7] Davies, J.C. 2006. Managing the effects of Nanotechnology. Woodrow Wilson Institute for Scholars. Available at www.wilsoncenter.org/nano; ICF International. 2006. Characterizing the environmental, health, and safety implications of nanotechnology: Where should the federal government go from here? ICF international, Fairfax, VA. Available at www.icfi.com; Maynard, A.D. 2006. Davies, J.C. 2006. Nanotechnology: A research strategy for addressing risk. Woodrow Wilson Institute for Scholars. Available at www.wilsoncenter.org/nano; Taylor, M.R. 2006. Regulating the products of nanotechnology: Does the FDA have the tools it needs? Woodrow Wilson Institute for Scholars. Available at www.wilsoncenter.org/nano .

[8] Chan, V.S.W. 2006. Nanomedicine: An Unresolved Regulatory Issue. *Regulatory Toxicology and Pharmacology*, 46 (3): 218-224; Davis, J.M. 2007. How to assess the risks of nanotechnology: Learning from past experience. *J. Nanoscience and Nanotechnology*, 7 1-8.; Oberdörster, G., et al. 2005. Nanotoxicology: an emerging discipline evolving from studies of ultrafine particles. *Environ. Health Perspect*. 113: 823-829.

[9] Tsuji JS et al. Research strategies for safety evaluation of nanomaterials, part IV: risk assessment of nanoparticles. *Toxicol Sci* 2006 89(1): 42-50; Oberdörster, G., et al., Principles for characterizing the potential human health effects from exposure to nanomaterials: elements of a screening strategy. *Part Fibre Toxicol* 2: 8 2005. http://ec.europa.eu/health/ph_risk/committees/04_scenihr/docs/scenihr_o_003b.pdf http://www.nanotec.org.uk/finalReport.htm

[10] Morgan, K. Development of a Preliminary Framework for Informing the Risk Analysis and Risk Management of Nanoparticles. *Risk Analysis*, 25(6): 1621-1635 2005.

[11] Warheit DB, et al. Pulmonary Bioassay Studies with Nanoscale and Fine-Quartz Particles in Rats: Toxicity is Not Dependent upon Particle Size but on Surface Characteristics. Toxicological Sciences 2007 95(1):270-280.; Warheit DB, et al. Comparative pulmonary toxicity inhalation and instillation studies with different TiO_2 particle formulations: Impact of surface treatments on particle toxicity. *Toxicological Sciences* 88(2): 514-524 2005.

[12] Hoshino A, et al. Physicochemical properties and cellular toxicity of nanocrystal quantum dots depend on their surface modification. *Nano Letters* 4(11): 2163-2169 NOV 2004; Warheit DB, et al. Pulmonary Bioassay Studies with Nanoscale and Fine-Quartz Particles in Rats: Toxicity is Not Dependent upon Particle Size but on Surface Characteristics. *Toxicological Sciences* 2007 95(1):270-280.; Warheit DB, et al. Comparative pulmonary toxicity inhalation and instillation studies with different TiO2 particle formulations: Impact of surface treatments on particle toxicity. *Toxicological Sciences* 88 (2): 514-524 2005.

[13] It is important to note with respect to the development of general knowledge that many of the studies published in the literature have been conducted with nanoscale materials that are either poorly characterized, or not characterized. Therefore, in many cases the quality and applicability of the findings reported may be inadequate.

[14] Lockman PR, et al. Nanoparticle surface charges alter blood-brain barrier integrity and permeability, *J. Drug Target.* 12(9-10): 635-641 (2004).

[15] Derfus AM, et al. Probing the cytotoxicity of semiconductor quantum dots, *Nano Lett.* 4(1):11-18 (2004).

[16] Ballou B, et al. Noninvasive imaging of quantum dots in mice, *Bioconjugate Chem.* 15(1): 79-86 (2004).

[17] Brown DM, Wilson MR, MacNee W, Stone V, Donaldson K. Size-dependent proinflammatory effects of ultrafine polystyrene particles: a role for surface area and oxidative stress in the enhanced activity of ultrafines. *Toxicol Appl Pharmacol* 2001, 175:191-199.

[18] Lynch I. 2007. Are there generic mechanisms governing interactions between nanoparticles and cells? Epitope mapping the outer layer of the proteinBmaterial interface. *Physica A: Statistical and Theoretical Physics* 373: 511-520.

[19] Description of the NTP's Nanotechnology Safety Initiative at: http://ntp.niehs.nih.gov/go/20837.

[20] FDA is currently reviewing this petition. FDA has not yet reached a decision on the petition because it raises complex issues requiring extensive review and analysis by agency officials, and in relation to which the agency is seeking public input. While this report addresses some issues raised in the petition, this report reflects only the views of the Task Force, and does not constitute an agency answer to the petition in whole or in part.

[21] National Nanotechnology Initiative (NNI). 2006. Environmental, Health, and Safety Research Needs for Engineered Nanoscale Materials. Available at www.nano.gov.

[22] http://www.nano.gov/NNI_EHS_research_needs.pdf; http://www.nanotec.org.uk/ final Report.htm

[23] Heinze, T. 2004. Nanoscience and nanotechnology in Europe: Analysis of publications and patent applications including comparisons with the United States. *Nanotechnology Law & Business* 1(4): 10.

[24] Beginning December 22, 2007, any serious adverse events associated with a dietary supplement reported to the product's manufacturer, packer, or distributor will have to be submitted to FDA. In addition, firms will have to keep records of all dietary supplement adverse events, serious or non-serious, and FDA will have access to those records during inspections.

[25] Teeguarden,J.G., Hinderliter, P.M., Orr, G., Thrall, B.D., and Pounds, J.G. 2007. Particokinetics in vitro: Dosimetry considerations for in vitro nanoparticle toxicity assessments. *Toxicol. Sciences* 95: 300-312.

[26] *M*inimum *I*nformation *A*bout a *M*icroarray *E*xperiment www.mged.org/ Workgroups/ MIAME/miame.html. Consistent reporting approaches can be established and instituted without adopting substantive definitions for "nanotechnology" and related terms for regulatory purposes (the merits of which may take some time to evaluate as explained above).

[27] For purposes of this report, the term "premarket authorization" is used to refer to a number of regulatory actions that the FFDCA and the PHS Act and agency regulations may refer to by other names, including "approval," "clearance," "licensing," and "listing." As used in this

report, "premarket authorization" includes both premarket approval for an individual product (e.g., under an NDA, BLA, or PMA) and regulations permitting the marketing or use of an ingredient or substance for specified uses under specified conditions (e.g., a food additive regulation, OTC drug monograph, or color additive listing regulation). Among other issues concerning the use of nanotechnology, this report focuses on how to treat nano versions of existing products that fall under such a regulation and makes recommendations to guide agency consideration of that issue.

[28] Generally, a drug is regulated as a "new" drug under section 505 of the FFDCA if it is not generally recognized to be safe and effective.

[29] A "new dietary ingredient" is a dietary ingredient that was not marketed in the United States before October 15, 1994. New dietary ingredients require a premarket notification to FDA unless the ingredient has been "present in the food supply as an article used for food in a form in which the food has not been chemically altered." 21 U.S.C. 350b.

[30] Notifications of a GRAS determination can also be submitted on a postmarket basis.

[31] Under section 351 of the PHS Act, FDA will approve a biologics license application on the basis of a demonstration that, among other things, the biological product that is the subject of the application is safe, pure, and potent. Potency has long been interpreted to include effectiveness. 21 CFR 600.3(s).

[32] A combination product is a product that is a drug-device, drug-biologic, device-biologic, or a drug-device-biologic. The assignment of a "lead center" that will have primary jurisdiction for its regulation is based on a determination of the "primary mode of action" of the combination product. FDA regulations define the primary mode of action of a combination product as "the single mode of action . . . that provides the most important therapeutic action of the combination product" (21 CFR 3.2(m)) and define therapeutic action or effect to include "any effect or action of the combination product intended to diagnose, cure, mitigate, treat, or prevent disease, or affect the structure or any function of the body" (21 CFR 3.2(k)).

[33] A biological product that is subject to licensure under section 351 of the Public Health Service Act (PHS Act) may meet the definition of "drug" under the FFDCA and thus be subject to certain provisions of drug regulation such as the IND regulations.

[34] Under certain circumstances, FDA does not object to the marketing of OTC drugs with active ingredients being considered for inclusion in an OTC monograph as part of the "OTC drug review."

[35] http://www.fda.gov/cdrh/ode/510kmod.pdf

[36] http://www.fda.gov/oc/ohrt/irbs/devrisk.pdf

[37] Although adverse events for color additives marketed as stand-alone products or used in conventional foods and cosmetics do not have to be reported to FDA, adverse events for color additives that are components of a drug or device must be reported as part of the adverse event reporting requirements for the finished product.

[38] However, a product that meets the definition for a cosmetic but is intended to affect the structure or function of the body of man will also be subject to regulation as a drug or device.

[39] A new dietary ingredient is an ingredient not marketed in the U.S. before October 15, 1994.

CHAPTER SOURCES

The following chapters have been previously published:

Chapter 1 – This is an edited, reformatted and augmented version of United States Government Accountability Office publication GAO-10-549, dated May 2010

Chapter 2 – This is an edited, reformatted and augmented version of Congressional Research Service publication, Report RL34614, dated February 9, 2009.

Chapter 3 – This is an edited, reformatted and augmented version of Nanotechnology Report United States Food and Drug Administration Nanotechnology Task Force, dated July 25 2007

INDEX

F

G

H

I

N

O

P

Q

R

S

T

U